Communications in Computer and Information Science 3078

Series Editors

Gang Li, *School of Information Technology, Deakin University, Burwood, VIC, Australia*
Joaquim Filipe, *Polytechnic Institute of Setúbal, Setúbal, Portugal*
Zhiwei Xu, *Chinese Academy of Sciences, Beijing, China*

AF183563

Rationale

The CCIS series is devoted to the publication of proceedings of computer science conferences. Its aim is to efficiently disseminate original research results in informatics in printed and electronic form. While the focus is on publication of peer-reviewed full papers presenting mature work, inclusion of reviewed short papers reporting on work in progress is welcome, too. Besides globally relevant meetings with internationally representative program committees guaranteeing a strict peer-reviewing and paper selection process, conferences run by societies or of high regional or national relevance are also considered for publication.

Topics

The topical scope of CCIS spans the entire spectrum of informatics ranging from foundational topics in the theory of computing to information and communications science and technology and a broad variety of interdisciplinary application fields.

Information for Volume Editors and Authors

Publication in CCIS is free of charge. No royalties are paid, however, we offer registered conference participants temporary free access to the online version of the conference proceedings on SpringerLink (http://link.springer.com) by means of an http referrer from the conference website and/or a number of complimentary printed copies, as specified in the official acceptance email of the event.

CCIS proceedings can be published in time for distribution at conferences or as post-proceedings, and delivered in the form of printed books and/or electronically as USBs and/or e-content licenses for accessing proceedings at SpringerLink. Furthermore, CCIS proceedings are included in the CCIS electronic book series hosted in the SpringerLink digital library at http://link.springer.com/bookseries/7899. Conferences publishing in CCIS are allowed to use our online conference service (Meteor) for managing the whole proceedings lifecycle (from submission and reviewing to preparing for publication) free of charge.

Publication process

The language of publication is exclusively English. Authors publishing in CCIS have to sign the Springer CCIS copyright transfer form, however, they are free to use their material published in CCIS for substantially changed, more elaborate subsequent publications elsewhere. For the preparation of the camera-ready papers/files, authors have to strictly adhere to the Springer CCIS Authors' Instructions and are strongly encouraged to use the CCIS LaTeX style files or templates.

Abstracting/Indexing

CCIS is abstracted/indexed in DBLP, Google Scholar, EI-Compendex, Mathematical Reviews, SCImago, Scopus. CCIS volumes are also submitted for the inclusion in ISI Proceedings.

How to start

To start the evaluation of your proposal for inclusion in the CCIS series, please send an e-mail to ccis@springer.com

Manuela Geiß · Ulrich Göhner ·
Sebnem Gül-Ficicir · Bernhard Moser ·
Gabriele Kotsis · A Min Tjoa · Ismail Khalil
Editors

Database and Expert Systems Applications - DEXA 2026 Workshops

37th International Conference on Database and Expert Systems Applications, DEXA 2026
Graz, Austria, August 11–13, 2026
Proceedings

Springer

Editors
Manuela Geiß
Software Competence Center Hagenberg
Hagenberg, Austria

Sebnem Gül-Ficicir
University of Applied Sciences Kempten
Kempten, Germany

Gabriele Kotsis
Johannes Kepler University Linz
Linz, Austria

Ismail Khalil
Johannes Kepler University Linz
Linz, Austria

Ulrich Göhner
University of Applied Sciences Kempten
Kempten, Germany

Bernhard Moser
Software Competence Center Hagenberg
Hagenberg, Austria

A Min Tjoa
Vienna University of Technology
Vienna, Austria

ISSN 1865-0929 ISSN 1865-0937 (electronic)
Communications in Computer and Information Science
ISBN 978-3-032-34201-0 ISBN 978-3-032-34202-7 (eBook)
https://doi.org/10.1007/978-3-032-34202-7

© The Editor(s) (if applicable) and The Author(s), under exclusive license
to Springer Nature Switzerland AG 2027

This work is subject to copyright. All rights are solely and exclusively licensed by the Publisher, whether the whole or part of the material is concerned, specifically the rights of translation, reprinting, reuse of illustrations, recitation, broadcasting, reproduction on microfilms or in any other physical way, and transmission or information storage and retrieval, electronic adaptation, computer software, or by similar or dissimilar methodology now known or hereafter developed.
The use of general descriptive names, registered names, trademarks, service marks, etc. in this publication does not imply, even in the absence of a specific statement, that such names are exempt from the relevant protective laws and regulations and therefore free for general use.
The publisher, the authors and the editors are safe to assume that the advice and information in this book are believed to be true and accurate at the date of publication. Neither the publisher nor the authors or the editors give a warranty, expressed or implied, with respect to the material contained herein or for any errors or omissions that may have been made. The publisher remains neutral with regard to jurisdictional claims in published maps and institutional affiliations.

This Springer imprint is published by the registered company Springer Nature Switzerland AG
The registered company address is: Gewerbestrasse 11, 6330 Cham, Switzerland

If disposing of this product, please recycle the paper.

Preface

Welcome to the Proceedings of the DEXA 2026 Workshops. This year, we again hosted two workshops: the 6th International Workshop on AI System Engineering: Math, Modelling and Software (AISys 2026) and the 2nd International Workshop on Optimisation of Industrial Production with AI Algorithms (AI4IP 2026). These events took place from August 11–13, 2026 in Graz, Austria.

This compilation of papers and presentations represents a convergence of cutting-edge research, interdisciplinary collaboration, industrial application of algorithms and methods, and innovative solutions at the forefront of science and technology.

AISys 2026 focused on making AI practical for real-world industrial and commercial applications. The contributions address the gap between powerful modern AI models and the constraints of applied settings—limited labeled data, opaque decision-making, heterogeneous data quality, and distributed infrastructure. Whether through few-shot learning, synthetic data generation, federated training, or explainability methods, each paper tackles a different barrier to deploying AI reliably and at scale outside of controlled research environments. Together, they reflect a shared ambition to bridge the divide between methodological advances and the demands of practical deployment.

AI4IP 2026 was initiated in 2025 by a group of scientists working in the field of AI algorithms applied to manufacturing processes. In modern production environments, the combination of artificial intelligence and advancing digitalization offers significant potential for increased productivity and more effective use of resources. Nevertheless, the application of smart technologies in practical industrial settings remains an active area of research. AI4IP 2026 brought together contributions addressing a central challenge in applied machine learning: the development and deployment of AI and machine learning methods for physical and cyber-physical systems. Whether simulating complex mechanical dynamics, fusing heterogeneous sensor modalities, reconstructing industrial communication infrastructures, or applying reinforcement learning to real-time control tasks, each contribution addresses the fundamental challenge of enabling machines to perceive, represent, and act within physically constrained environments. The workshop also covered the retrieval of production-relevant information from technical drawings, demonstrating that traditional engineering artifacts are increasingly being integrated into fully automated manufacturing workflows. The papers reflect a principled effort to embed data-driven AI methods—including reinforcement learning, federated learning, physics-informed neural networks, and self-supervised learning—into domains where physical laws, operational safety requirements, and sensor limitations impose stringent real-world constraints.

Collectively, the contributions reflect a sustained ambition to develop intelligent, adaptive systems capable of operating reliably under the complex and heterogeneous conditions that characterize modern industrial environments. The two workshops attracted

29 submissions, from which only 14 papers were accepted after a rigorous single-blind peer-review process in which submissions received on average three reviews each, ensuring the highest standards of quality and relevance to the DEXA theme.

We extend our deepest gratitude to the authors, reviewers, and organizers whose dedication and expertise made these workshops possible. Their collective efforts ensured that the knowledge shared here is of the highest quality and relevance.

As you engage with these proceedings, we hope you find inspiration, fresh ideas, and valuable insights that will enrich your work and contribute meaningfully to the ongoing advancement of this dynamic and impactful field.

August 2026

Manuela Geiß
Ulrich Göhner
Sebnem Gül-Ficicir
Bernhard Moser
Gabriele Kotsis
A Min Tjoa
Ismail Khalil

Organization

Steering Committee

Gabriele Kotsis	Johannes Kepler University Linz, Austria
A Min Tjoa	Vienna University of Technology, Austria
Manuela Geiß	Software Competence Center Hagenberg, Austria
Bernhard Moser	Software Competence Center Hagenberg/Institute of Signal Processing, Johannes Kepler University Linz, Austria
Christine Strauss	University of Vienna, Austria
Ismail Khalil	Johannes Kepler University Linz, Austria

AISys 2026 Chairs

Manuela Geiß	Software Competence Center Hagenberg, Austria
Bernhard Moser	Software Competence Center Hagenberg, Austria

AISys 2026 Program Committee

Lukas Fischer	NXAI, Austria
Michael Granitzer	University of Passau, Germany
Bernhard Heinzl	Software Competence Center Hagenberg, Austria
Maqbool Khan	American University of Bahrain, Bahrain
Elmar Kiesling	Institute for Data, Process and Knowledge Management at WU Vienna, Austria
Franz Krause	University of Mannheim, Germany
Jorge Martinez-Gil	Software Competence Center Hagenberg, Austria
Georg Thallinger	Joanneum Research, Austria
Wolfram Wöß	Johannes Kepler University Linz, Austria

AI4IP 2026 Chairs

Uli Göhner	University of Applied Sciences Kempten, Germany
Sebnem Gül-Ficici	University of Applied Sciences Kempten, Germany
Dirk Jacob	University of Applied Sciences Kempten, Germany
Frank Schirmeier	University of Applied Sciences Kempten, Germany

AI4IP 2026 Program Committee

Anton Dignös	Free University of Bozen-Bolzano, Italy
Sebastian Klüpfel	Bosch GmbH, Germany
Stefan Huber	Salzburg University of Applied Sciences, Austria

Organizers

scch { }
software competence
center hagenberg

JKU
Institute for
Telecooperation

Web Applications Society
@WAS
www.iiwas.org

Contents

AI System Engineering: Math, Modelling and Software

Physics-Informed Estimation of Behind-the-Meter PV Self-consumption with Battery-Aware Monte Carlo Simulation

Marius Führer(✉)

FH Technikum Wien, Vienna, Austria
marius.fuehrer@technikum-wien.at

Abstract. Estimating photovoltaic (PV) self-consumption behind the meter is critical for evaluating distributed solar-plus-storage systems, yet standard smart meters record only net grid flows. This paper presents a physics-informed Monte Carlo framework, developed within the FAIR-AI project (https://fair-ai.at), that reconstructs unobserved PV generation from the pvlib Ineichen clear-sky model, derives a consumer-load-corrected cloud attenuation factor, and simulates plausible battery dispatch under dual demand proxies to produce calibrated uncertainty bounds on self-consumption. Validated on a synthetic dataset of 50 prosumers with known ground truth, the method achieves 86% P10–P90 coverage (MAE = 0.131), and bootstrap aggregation tightens the community-level uncertainty band by a factor of 5.5× (from 0.472 to 0.086). Applied to 124 real Austrian prosumers over 22 months, the community-level self-consumption rate is 49.1% (bootstrap P10–P90: [41.0%, 55.9%]).

Keywords: Photovoltaic Self-Consumption · Behind-the-meter Estimation · Physics-Informed Modelling · Monte Carlo Simulation · Battery Energy Storage · Smart Meter Data

1 Introduction

As feed-in tariffs decline across Europe [4], the economics of rooftop PV increasingly hinge on the self-consumption rate (SCR), the fraction of generation consumed on-site. Battery energy storage systems raise the SCR by time-shifting surplus PV [11,18], but they also decouple measured export from actual generation: standard smart meters record only net import and export, while PV output, battery state, and on-site demand remain hidden behind the meter (BTM).

Existing approaches to SCR estimation have significant limitations. Direct metering requires dedicated PV sensors not universally deployed. Statistical disaggregation [5,10] requires labelled training data. Recent BTM disaggregation methods [1,15,16] typically need auxiliary measurements, while physics-based

© The Author(s), under exclusive license to Springer Nature Switzerland AG 2027
M. Geiß et al. (Eds.): DEXA 2026 Workshops, CCIS 3078, pp. 3–14, 2027.
https://doi.org/10.1007/978-3-032-34202-7_1

modelling [8, 19] ignores storage, leading to systematic bias when batteries are present. Supervised neural and deep-learning disaggregation approaches could in principle absorb the storage-induced bias, but they require labelled behind-the-meter ground truth (per-user PV, demand, and battery traces) that is unavailable in the deployment setting considered here; we therefore do not benchmark against them directly.

We address this gap with a *physics-informed, battery-aware* Monte Carlo framework whose key contributions are:

1. PV generation reconstruction using the pvlib Ineichen clear-sky model with a consumer-load-corrected cloud factor, requiring *only* smart meter data and site coordinates.
2. Dual demand proxies—a BTM-based upper bound and an import-profile lower bound—combined with a three-category battery prior in a joint Monte Carlo simulation.
3. Bootstrap aggregation that reduces community-level uncertainty by a factor of 5.5 relative to per-user bounds.
4. Comprehensive validation on synthetic data with known ground truth, and application to 124 real Austrian prosumers.

This work extends the PV self-consumption module of a broader FAIR-AI pipeline for energy community analytics [3], which additionally includes explainable consumption forecasting and battery storage optimisation.

2 Data

Two datasets are used: a synthetic dataset with known ground truth for validation, and a real-world dataset for practical application. The synthetic dataset contains 50 prosumers (100 metering points) at 15-min resolution throughout 2024 ($\sim$3.5 M observations), with ground truth for each user including true PV capacity, battery parameters, and true SCR. The real dataset comprises 124 residential prosumers in an Austrian energy community, spanning January 2024 to October 2025 ($\sim$4.76 M observations). Each record contains a user identifier, timestamp, grid export, and grid import; no ground truth is available. Timestamps are local civil time (Europe/Vienna, DST retained); export and import are interval energies (kWh per 15-min slot), converted to power by dividing by $\Delta t = 0.25\,\text{h}$. A user is not expected to import or export at every interval, so empty intervals are not treated as missing data; no systematic gaps were observed at coarser scales. Table 1 summarises both datasets.

3 Methodology

3.1 The Non-identifiability Problem

At any interval t, the meter records grid import $I_i(t) \geq 0$ and grid export $E_i(t) \geq 0$. The true energy balance for prosumer i is:

$$G_i(t) + I_i(t) = D_i(t) + E_i(t) + \Delta B_i(t), \tag{1}$$

Table 1. Dataset summary.

Parameter	Synthetic	Real
Prosumers	50	124
Period	Jan–Dec 2024	Jan 2024–Oct 2025
Resolution	15 min	15 min
Observations	~3.5 M	~4.76 M
Ground truth	Yes	No

where $G_i(t)$ is true PV generation, $D_i(t)$ is true demand, and $\Delta B_i(t)$ is net battery charging. Since only I_i and E_i are observed, generation, demand, and battery state are jointly unidentifiable. Our approach uses physical constraints and Monte Carlo sampling to produce calibrated uncertainty bounds rather than point estimates.

3.2 Clear-Sky Irradiance Model

We use the pvlib library's Ineichen–Perez clear-sky model [6,7] with climatological Linke turbidity [13]. For each timestamp, pvlib computes the solar position via the NREL SPA algorithm [12], looks up the monthly Linke turbidity factor T_L for the site coordinates, and evaluates the Ineichen broadband clear-sky GHI:

$$G_{\text{cs}}(t) = a_1 \cdot I_0 \sin(\alpha_s) \cdot \exp\!\big(-a_2 \cdot \text{AM} \cdot (f_{h1} + f_{h2}(T_L - 1))\big) \cdot \exp(0.01 \cdot \text{AM}^{1.8}), \quad (2)$$

where $I_0 = 1361\ \text{W/m}^2$, α_s is solar elevation, AM is air mass, and a_1, a_2, f_{h1}, f_{h2} are altitude-dependent coefficients [14]. We set $G_{\text{cs}} = 0$ when $\alpha_s \leq 0$ and normalise to $f_{\text{cs}}(t) = G_{\text{cs}}(t)/\max_t G_{\text{cs}}(t) \in [0, 1]$. For the study site, the maximum clear-sky GHI is $866\,\text{W/m}^2$.

3.3 Cloud Attenuation with Load Correction

Under clear skies, the community-aggregate export tracks f_{cs} closely. We estimate the community PV capacity as the 99th percentile of the export-to-irradiance ratio during daylight:

$$\hat{P}_{\text{comm}} = Q_{0.99}\left(\frac{E_{\text{comm}}(t)}{f_{\text{cs}}(t)}\right), \quad f_{\text{cs}}(t) > 0.10, \quad (3)$$

yielding a base effective irradiance $f_{\text{eff,base}}(t) = E_{\text{comm}}(t)/\hat{P}_{\text{comm}}$.

However, $f_{\text{eff,base}}$ systematically underestimates true irradiance during high-demand periods, because self-consumption reduces export even when irradiance is unchanged. To correct this, we build a normalised consumer load profile $\lambda(m, h, w) \in [0, 1]$ from periods with low community export (below the 50th percentile), stratified by month m, hour h, and weekday/weekend w:

$$\lambda(m,h,w) = \frac{\text{median}_{t\in\mathcal{T}_{\text{low}}}(I_{\text{comm}}(t) \mid m,h,w)}{\max_{m,h,w}(\cdot)}. \tag{4}$$

The corrected effective irradiance incorporates the load correction with tunable strength $\alpha \geq 0$:

$$f_{\text{eff}}(t) = \min\big(f_{\text{cs}}(t),\ f_{\text{eff,base}}(t)\cdot(1+\alpha\cdot\lambda(t))\big), \tag{5}$$

clipped to $[0,1]$. The min enforces the physical constraint that effective irradiance cannot exceed clear-sky. Section 4.2 presents a sensitivity analysis over α; we adopt $\alpha = 0.5$.

3.4 Per-User PV Capacity and Generation Estimate

For each prosumer i, the capacity proxy is estimated as the 99.5th percentile of $E_i(t)/f_{\text{eff}}(t)$ over daylight intervals with positive export (requiring ≥ 40 qualifying intervals; fallback: community median). The PV estimate is:

$$\hat{G}_i(t) = \max\big(\hat{P}_i \cdot f_{\text{eff}}(t),\ E_i(t)\big), \tag{6}$$

where the max enforces that generation must be at least as large as observed export.

3.5 Dual Demand Proxies

Since true demand is unobserved during PV hours, we construct two complementary proxies.

The *BTM-based proxy* (upper bound) uses the energy balance:

$$\hat{D}_i^{\text{BTM}}(t) = I_i(t) + \max\big(\hat{G}_i(t) - E_i(t),\ 0\big), \tag{7}$$

attributing all non-exported generation to consumption (overestimating demand when a battery absorbs surplus).

The *import-profile proxy* (lower bound) builds a per-user demand profile from nighttime and low-irradiance periods ($f_{\text{cs}} < 0.20$ or zero export), stratified by month and hour:

$$\hat{D}_i^{\text{prof}}(m,h) = \text{median}_{t\in\mathcal{T}_{\text{rel}}}(I_i(t) \mid m,h). \tag{8}$$

During PV hours ($f_{\text{cs}} > 0.05$) this is floored by the observed import,

$$\max\big(\hat{D}_i^{\text{prof}}(t),\ I_i(t)\big),$$

and reverts to $I_i(t)$ otherwise. This proxy underestimates daytime demand because it extrapolates nighttime patterns.

3.6 Battery-Aware Monte Carlo Simulation

We run $N = 100$ MC trials per user, each jointly sampling a demand proxy (BTM or profile, equal weight) and a battery configuration. The 50/50 weighting is the maximum-entropy choice given the symmetric bracketing of Sect. 3.5—an upper and a lower bound on true daytime demand—and is what makes the ensemble calibrated; auxiliary demand information could replace the uniform mixture with an informed weight. The battery configuration is drawn from

$$\text{Battery} \sim \begin{cases} \text{None:} & S_{\max} = P_{\max} = 0, \quad \text{w.p. } \frac{1}{3}, \\ \text{Small:} & S_{\max} \sim \mathcal{U}(3,8)\,\text{kWh},\ P_{\max} \sim \mathcal{U}(1.5,3)\,\text{kW}, \quad \text{w.p. } \frac{1}{3}, \\ \text{Large:} & S_{\max} \sim \mathcal{U}(8,15)\,\text{kWh},\ P_{\max} \sim \mathcal{U}(3,5)\,\text{kW}, \quad \text{w.p. } \frac{1}{3}. \end{cases} \tag{9}$$

The equal 1/3 weighting is again a maximum-entropy stance: in the absence of installation records the prior brackets the residential segments documented in [17,18] (no storage, ~5 kWh small, ~10–15 kWh large) without concentrating on one class. Its coarseness inflates per-user bounds for battery-equipped users, but bootstrap aggregation (Sect. 3.7) absorbs this mismatch at the community scale; deployments with known battery penetration can substitute a calibrated mixture. Initial SOC is drawn from $\mathcal{U}(0.2, 0.8)$; charge and discharge efficiencies are $\eta_c = \eta_d = 0.95$.

Greedy Dispatch. At each 15-min interval ($\Delta t = 0.25$ h), let $G(t)$ and $L(t)$ denote the PV and demand estimates, $C_{\max} = D_{\max} = P_{\max}\,\Delta t$, and $s(t) = \max(G(t) - L(t),\, 0)$ the PV surplus. A self-consumption-maximising rule [17,18] first matches PV to load, $\text{SC}_{\text{direct}}(t) = \min(G(t), L(t))$, then charges the battery with the surplus and discharges to cover any residual load:

$$\text{ch}(t) = \min\bigl(s(t),\, C_{\max},\, (S_{\max} - \text{SOC}(t))/\eta_c\bigr), \tag{10}$$

$$\text{SOC}(t^{+}) = \text{SOC}(t) + \eta_c\,\text{ch}(t), \tag{11}$$

$$\text{dis}(t) = \min\bigl(L(t) - \text{SC}_{\text{direct}}(t),\, D_{\max},\, \text{SOC}(t^{+})\,\eta_d\bigr), \tag{12}$$

$$\text{SOC}(t^{++}) = \text{SOC}(t^{+}) - \text{dis}(t)/\eta_d. \tag{13}$$

The per-user SC rate for trial k is $\text{SCR}_{i,k} = \sum_t[\text{SC}_{\text{direct},k}(t) + \text{dis}_k(t)] / \sum_t \hat{G}_i(t)$, and quantiles across trials yield per-user bounds (P10, P50, P90).

3.7 Bootstrap Aggregation

Per-user bounds are wide because each trial samples an entire battery configuration; community-level aggregation tightens them via the law of large numbers. The community SC rate per trial is $\text{SCR}_{\text{comm},k} = \sum_i \text{SC}_{i,k} / \sum_i \hat{G}_i$. To capture uncertainty from community composition we perform $B = 2000$ bootstrap resamples (draw n users with replacement, assign each a random MC trial, compute the PV-weighted community SCR), yielding the P10–P90 interval and 95% CI. For the hourly profile we run $N_{\text{TS}} = 100$ per-timestep trials with a community-level demand mixing weight $w_k \sim \mathcal{U}(0,1)$, $\hat{D}_i(t) =$

$w_k\,\hat{D}_i^{\mathrm{BTM}}(t) + (1 - w_k)\,\hat{D}_i^{\mathrm{prof}}(t)$, and independent per-user battery draws so that per-user noise averages out at the community scale.

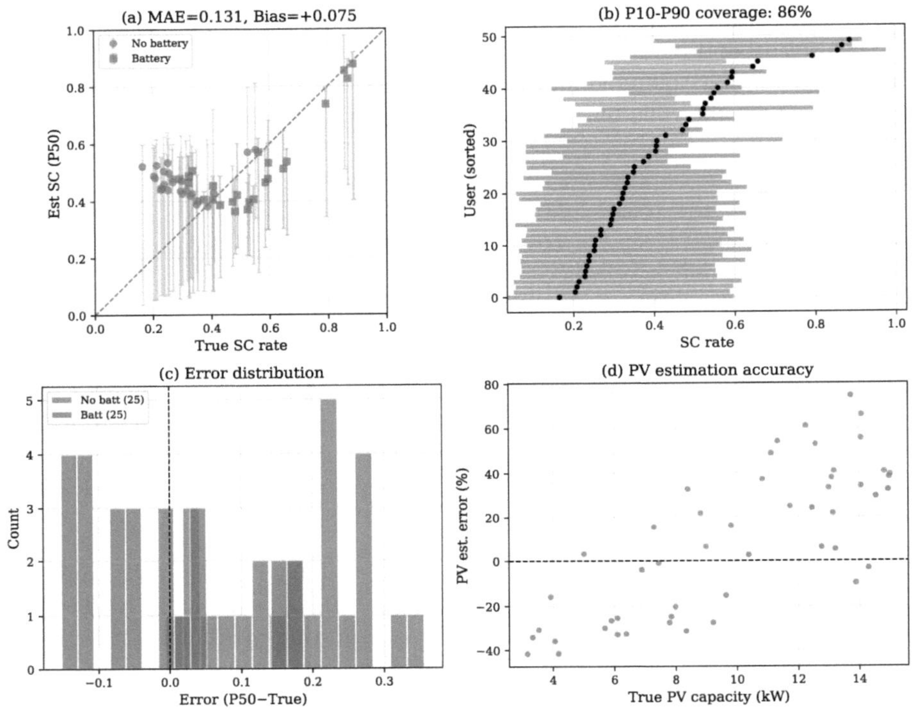

Fig. 1. Synthetic validation ($n = 50$). (a) P50 vs. true SCR with P10–P90 bars. (b) Per-user coverage (green = in bounds). (c) Error distribution by battery status. (d) PV estimation error vs. true capacity. (Color figure online)

4 Results

4.1 Synthetic Data Validation

Figure 1 summarises the per-user validation on the 50-prosumer synthetic dataset. The overall MAE is 0.131 with a positive bias of +0.075, largely attributable to the BTM demand proxy overstating demand for users without batteries. The P10–P90 coverage is 86% (43/50 users) and the P05–P95 coverage is 92%, both close to their nominal levels (80% and 90%), indicating well-calibrated tail bounds. The P25–P75 coverage of 82%, however, substantially exceeds the nominal 50%, revealing that the central interval is conservative: the MC ensemble over-disperses around the median, likely because the uniform battery prior assigns equal weight to configurations far from any individual user's true setup. This is a deliberate design choice—wide per-user bounds are acceptable when the primary target is the community-level aggregate, where bootstrap averaging compresses the band (Sect. 4.3).

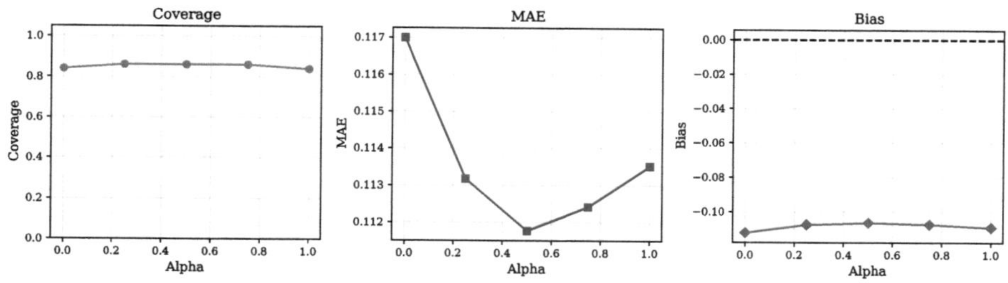

Fig. 2. Sensitivity to α: (a) P10–P90 coverage, (b) MAE, (c) bias.

4.2 Sensitivity to α

Figure 2 shows coverage, MAE, and bias as functions of the load-correction strength $\alpha \in \{0, 0.25, 0.5, 0.75, 1.0\}$. Coverage remains above 80% across the entire range, peaking at $\alpha = 0.5$, which also minimises MAE. Bias is mildly negative throughout, confirming that the method is robust to the choice of α within typical operating ranges.

4.3 Community-Level Aggregation

Figure 3 demonstrates the effect of aggregation. The true community SCR of 0.358 falls within the bootstrap P10–P90 band [0.324, 0.410], with a P50 estimate of 0.367 (error: +0.009) while the estimator for the simple aggregation is 0.465 (error: +0.107). Bootstrap aggregation reduces the P10–P90 width from a per-user median of 0.472 to a community-level 0.086—a 5.5× reduction, demonstrating that while individual SCRs are highly uncertain, the community-level rate is well-constrained. Sub-group validation by PV size tercile confirms that the true SCR falls within estimated bounds for all sub-groups.

4.4 Hourly Self-consumption Profile

Figure 4 shows the hourly SC profile validated against reconstructed ground truth. The ground truth falls within the estimated P10–P90 band for 77% of all solar hours and 100% of core solar hours (> 5% of peak PV). The profile shape is well-captured: SC rates are highest in early morning and late afternoon (when PV is small relative to demand) and lowest at midday. Over core solar hours the MAE is 0.116 and the bias is −0.048; residual errors concentrate at the edges of the solar window, where stored energy from the battery is discharged against a near-zero PV denominator and the ratio metric saturates at 1.

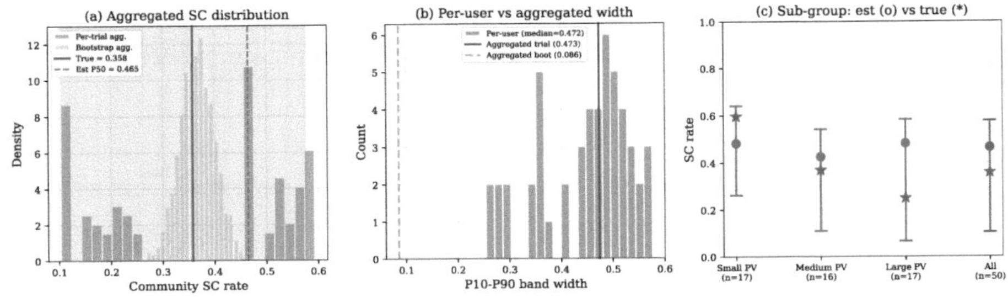

Fig. 3. Community-level aggregation. (a) Per-trial and bootstrap SC distributions with true value (red). (b) Per-user vs. aggregated band width. (c) Sub-group validation: estimated (o) vs. true (*). (Color figure online)

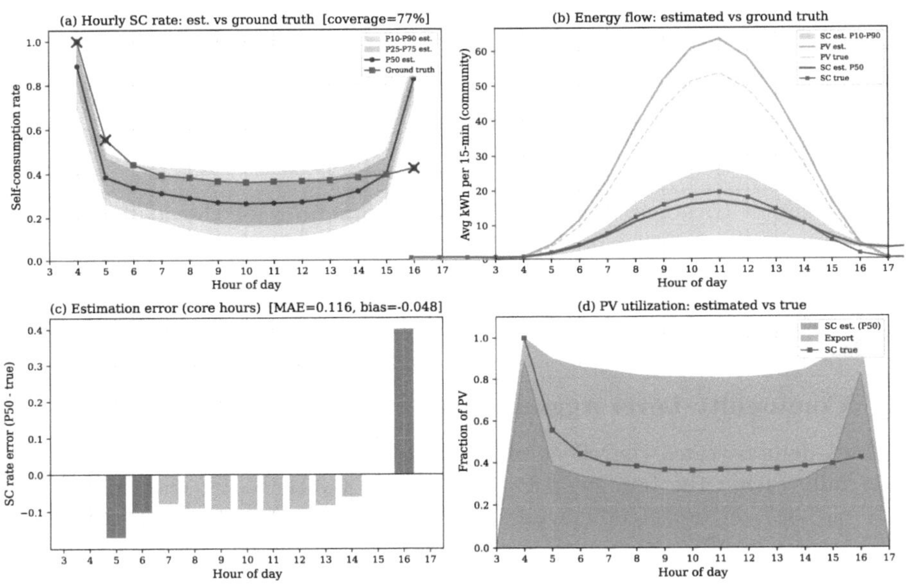

Fig. 4. Hourly SC profile—synthetic validation. (a) Estimated bands with ground truth (red). (b) Energy flows. (c) Per-hour error. (d) PV utilisation. (Color figure online)

4.5 Real Data Application

Applied to 124 real Austrian prosumers (Fig. 5), the median per-user P50 SC rate is 0.756 (P10–P90 width 0.723, reflecting the inherent ambiguity without ground truth), and bootstrap aggregation yields a community SC rate of 0.491 (P10–P90: [0.410, 0.559]), consistent with typical European residential SCRs for PV-plus-storage [11, 18]. No measured PV, demand, or battery state is available for this cohort, so these are plausible estimates with calibrated uncertainty rather than externally validated measurements; validation against a metered sub-sample is left to future work. The hourly profile (Fig. 6) shows the expected

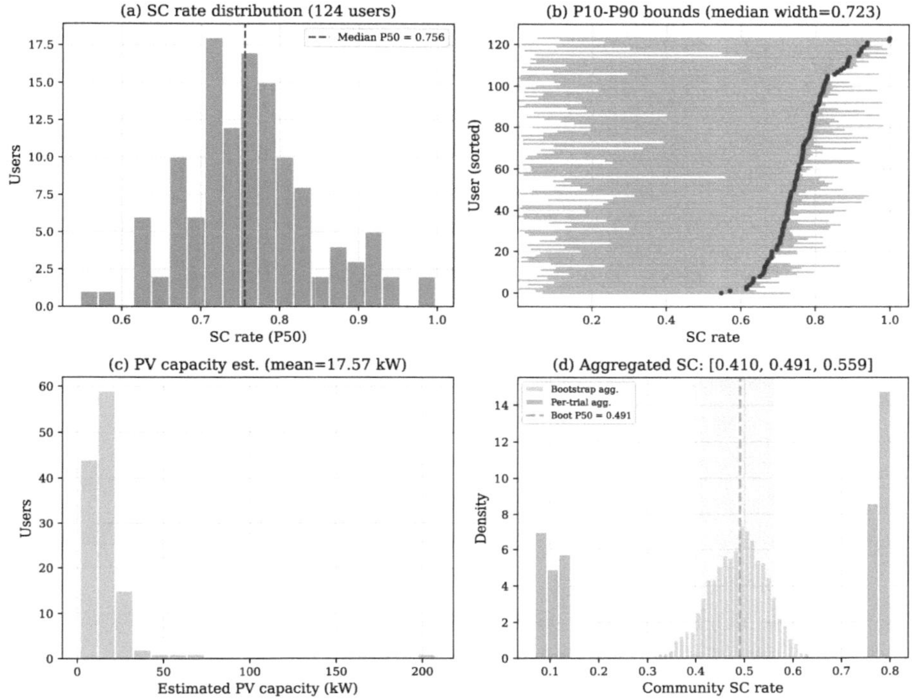

Fig. 5. Real data ($n = 124$). (a) Per-user SC distribution. (b) P10–P90 intervals sorted by P50. (c) PV capacity estimates. (d) Bootstrap community SC.

pattern: near-unity rates at solar edges declining to a midday minimum, with energy flows consistent with the synthetic validation.

5 Discussion

The pvlib Ineichen clear-sky model [6, 7], parameterised only by site coordinates and climatological Linke turbidity, makes the framework portable to other temperate residential settings without site-specific calibration; transferability to substantially different climates or installations would need empirical verification. The consumer-load correction (Eq. 5) removes a bias inherent to export-derived cloud factors—during high-demand periods, self-consumption suppresses export independently of cloud cover—and the sensitivity analysis confirms robustness across $\alpha \in [0, 1]$. The dual demand proxy design produces calibrated bounds by bracketing the unobservable true demand between an upper (BTM) and a lower (import-profile) bound; bootstrap aggregation then compresses the community-level band by 5.5×, making the method particularly suitable for operators needing aggregate SCR estimates.

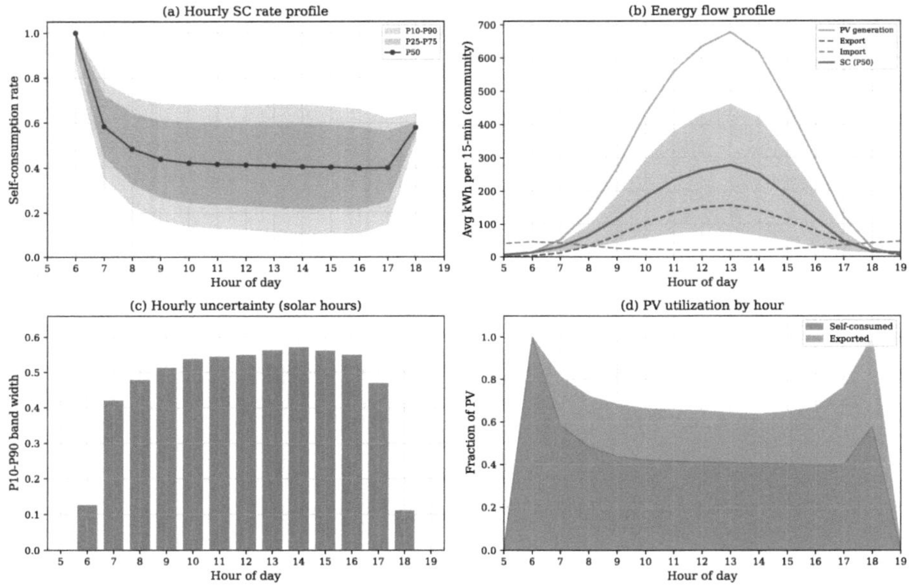

Fig. 6. Hourly SC profile—real data. (a) Estimated SC rate with P10–P90 bands. (b) Energy flows. (c) Uncertainty width. (d) PV utilisation.

Failure Modes. Accuracy degrades beyond the calibrated bounds when (i) cloud cover is spatially non-uniform across the community, breaking the aggregate cloud-factor assumption; (ii) site coordinates are inaccurate, biasing the solar-position and clear-sky reconstruction; (iii) the per-user capacity proxy is distorted by heavy over-consumption or partial shading (under-estimate) or by curtailment or clipping (over-estimate); or (iv) real dispatch deviates from greedy self-consumption (ToU arbitrage, grid services, scheduled charging). The deepest limitation is that battery state is non-identifiable from net-meter data alone (Eq. 1), so the method marginalises over battery existence rather than inferring it—the resulting wide per-user bounds reflect physical uncertainty, not estimation error. The 13% ground-truth SC error for battery users in the synthetic study is similarly an artefact of reconstructing demand from Eq. 1 when batteries shift energy across intervals, not of the estimator.

6 Conclusion

We have presented a physics-informed Monte Carlo framework for estimating BTM PV self-consumption from smart meter data. The method reconstructs PV generation using the pvlib Ineichen clear-sky model with consumer-load-corrected cloud attenuation, and applies battery-aware simulation with dual demand proxies to produce calibrated uncertainty bounds. Validated on synthetic data, it achieves 86% P10–P90 coverage (MAE = 0.131) and yields a community-level bootstrap band of width 0.086 around the true SCR. Applied to 124 real Austrian prosumers, the community SCR is estimated at 49.1% (P10–P90: [41.0%, 55.9%]); without ground truth this should be read as a plausible estimate with calibrated uncertainty rather than a validated measurement. The approach requires no external irradiance products, no PV metering, and no labelled site-level training data, making it applicable for energy community operators [2, 9] and distribution system operators without bespoke instrumentation. In combination with the consumption forecasting and battery optimisation modules of the wider pipeline [3], the self-consumption estimates can directly inform community-level storage planning. Nevertheless will it be crucial to test the method on real-world ground truth to identify possible adjustments.

Acknowledgments. This work was conducted within the FAIR-AI flagship project (https://fair-ai.at), funded by the Austrian Research Promotion Agency (FFG) under the AI Mission Austria (AIM AT) programme, FFG grant no. 904624.

Disclosure of Interests. The author has no competing interests to declare that are relevant to the content of this article.

References

1. Brown, J., Abate, A., Rogers, A.: Disaggregation of household solar energy generation using censored smart meter data. Energy Build. **231**, 110617 (2021). https://doi.org/10.1016/j.enbuild.2020.110617
2. Caramizaru, A., Uihlein, A.: Energy communities: an overview of energy and social innovation. Tech. report EUR 30083 EN, Publications Office of the European Union, Joint Research Centre (2020). https://doi.org/10.2760/180576
3. Führer, M., et al.: Explainable data-driven smart meter analysis. In: Proceedings of the 19th Forschungsforum der österreichischen Fachhochschulen (FFH), Graz, Austria (2026, to appear)
4. Günther, C., Schill, W.P., Zerrahn, A.: Prosumage of solar electricity: tariff design, capacity investments, and power sector effects. Energy Policy **152**, 112168 (2021). https://doi.org/10.1016/j.enpol.2021.112168
5. Hart, G.W.: Nonintrusive appliance load monitoring. Proc. IEEE **80**(12), 1870–1891 (1992). https://doi.org/10.1109/5.192069
6. Holmgren, W.F., Hansen, C.W., Mikofski, M.A.: pvlib python: a python package for modeling solar energy systems. J. Open Source Softw. **3**(29), 884 (2018). https://doi.org/10.21105/joss.00884
7. Ineichen, P., Perez, R.: A new airmass independent formulation for the Linke turbidity coefficient. Sol. Energy **73**(3), 151–157 (2002). https://doi.org/10.1016/S0038-092X(02)00045-2
8. Killinger, S., Engerer, N., Müller, B.: QCPV: a quality control algorithm for distributed photovoltaic array power output. Sol. Energy **143**, 120–131 (2017). https://doi.org/10.1016/j.solener.2016.12.053
9. Lowitzsch, J., Hoicka, C.E., van Tulder, F.J.: Renewable energy communities under the 2019 European Clean Energy Package - governance model for the energy clusters of the future? Renew. Sustain. Energy Rev. **122**, 109489 (2020). https://doi.org/10.1016/j.rser.2019.109489

10. Luthander, R., Widén, J., Munkhammar, J., Lingfors, D.: Self-consumption enhancement and peak shaving of residential photovoltaics using storage and curtailment. Energy **112**, 221–231 (2016). https://doi.org/10.1016/j.energy.2016.06.039
11. Luthander, R., Widén, J., Nilsson, D., Palm, J.: Photovoltaic self-consumption in buildings: a review. Appl. Energy **142**, 80–94 (2015). https://doi.org/10.1016/j.apenergy.2014.12.028
12. Reda, I., Andreas, A.: Solar position algorithm for solar radiation applications. Sol. Energy **76**(5), 577–589 (2004). https://doi.org/10.1016/j.solener.2003.12.003
13. Remund, J., Wald, L., Lefèvre, M., Ranchin, T., Page, J.: Worldwide Linke turbidity information. In: Proceedings of the ISES Solar World Congress, Göteborg, Sweden (2003)
14. Reno, M.J., Hansen, C.W., Stein, J.S.: Global horizontal irradiance clear sky models: implementation and analysis. Technical report. SAND2012-2389, Sandia National Laboratories (2012). https://doi.org/10.2172/1039404
15. Shaker, H., Zareipour, H., Wood, D.: Estimating power generation of invisible solar sites using publicly available data. IEEE Trans. Smart Grid **7**(5), 2456–2465 (2016). https://doi.org/10.1109/TSG.2016.2533164
16. Sossan, F., Nespoli, L., Medici, V., Paolone, M.: Unsupervised disaggregation of photovoltaic production from composite power flow measurements of heterogeneous prosumers. IEEE Trans. Industr. Inf. **14**(9), 3904–3913 (2018). https://doi.org/10.1109/TII.2018.2791932
17. Truong, C.N., Naumann, M., Karl, R.C., Müller, M., Jossen, A., Hesse, H.C.: Economics of residential photovoltaic battery systems in Germany: The case of Tesla's Powerwall. Batteries **2**(2), 14 (2016). https://doi.org/10.3390/batteries2020014
18. Weniger, J., Tjaden, T., Quaschning, V.: Sizing of residential PV battery systems. Energy Procedia **46**, 78–87 (2014). https://doi.org/10.1016/j.egypro.2014.01.160
19. Widén, J., Wäckelgård, E., Paatero, J., Lund, P.: Impacts of distributed photovoltaics on network voltages: stochastic simulations of three Swedish low-voltage distribution grids. Electric Power Syst. Res. **80**(12), 1562–1571 (2010). https://doi.org/10.1016/j.epsr.2010.07.007

Visual Explanations for Segmenting Objects for Tunnel Monitoring from Few Samples

Helmut Neuschmied[1(✉)], Martin Winter[1], Werner Bailer[1], Georg Thallinger[1], Peter Steiner[2], Jan Pfeiffer[2], and Michael Mett[2]

[1] JOANNEUM RESEARCH – DIGITAL, Graz, Austria
helmut.neuschmied@joanneum.at
[2] dibit Messtechnik GmbH, Innsbruck, Austria

Abstract. Ongoing monitoring of tunnel infrastructure is critical for safety but currently suffers from the high cost of manual inspection and the scarcity of labeled data for automated systems. We present a workflow for instance segmentation of 2D tunnel orthophotos using few-shot object detection (FSOD) to identify infrastructure elements and damage. Our approach leverages the DINOv3-enhanced DE-ViT framework and the domain-adaptive CD-ViTO detector, which utilizes self-supervised features to learn from limited samples. Central to our work is the use of explanations based on patch-level visualizations and 2D embedding analysis to identify noisy training patches and non-representative outliers. Evaluation on a rail tunnel dataset shows that domain adaptation significantly enhances performance, and that optimizations driven by insights from XAI further improve the accuracy of the resulting class prototypes.

Keywords: visual explanations · few-shot object detection · DINO

1 Introduction

We address the use case of ongoing monitoring of road an rail tunnels, in order to assess their condition and detect damages. The process starts from orthorectified and georeferenced true color images retrieved from a Dibit tunnel scanning system[1]. The Dibit8 software manages detected features such as tunnel information system (TIS) objects interacting with a 3D Tunnel model. Multi-temporally available tunnel scans and TIS objects offer the unique opportunity to quantify changing tunnel surfaces. By comparing automated and reproducible TIS objects of different time epochs, the formation and alteration of damage zones can easily be observed (e.g., crack growth, changing sintering areas, etc.). An example of a 2D orthophoto from a scan is shown in Fig. 1.

[1] https://www.dibit.at/en/solutions/.

© The Author(s), under exclusive license to Springer Nature Switzerland AG 2027
M. Geiß et al. (Eds.): DEXA 2026 Workshops, CCIS 3078, pp. 15–26, 2027.
https://doi.org/10.1007/978-3-032-34202-7_2

In order to generate TIS objects as far as possible from the obtained images, instance segmentation is required, as certain applications demand objects to be separated on a precise point or pixel resolution. Specific tunnel objects are difficult or impossible to detect using currently applied detection methods. Such objects are typically characterized by an elevated position in relation to the tunnel profile or a large extent (e.g., railway tracks, cables, lines, signs etc.). Depending on the 3D scanning system and 3D reconstruction method, such elements are prone to be neglected or erroneously projected on resulting orthorectified tunnel maps. Erroneous projections can cause incorrect localization of important tunnel infrastructure elements. At the same time these areas could conceal areas of critical damage that could then not be assessed during tunnel inspection. Fully segmented and labelled tunnels represent the data basis for extensive monitoring tasks and the creation of Building Information Models (BIM) offering a fundamental planning platform not only for planning and constructing new tunnels but also for efficient maintenance of existing tunnel structures. To assess the structural health status of distinct parts of a tunnel (e.g. block-wise), results from a damage segmentation and object segmentation need to be combined. However, highly precise ground truth annotations of the objects of interest as input for training are scarce, and expensive to produce. In addition, quality assessment of segmentation results and human-aided correction of falsely detected objects or erroneously segmented cracks require a lot of effort.

Overall, the aim is to develop and provide a robust and efficient (instance) segmentation approach to label the 3D point cloud data. For many tasks, 2D segmentation approaches are still more robust [9,12], so that we follow an approach of segmenting 2D data and projecting labels to the point cloud via the known registration of images. The following challenges exist:

Few Labelled Samples. Labelling data is labour intensive and thus costly. Thus the aim is to learn object classes from as few samples as possible. In such a case an initial model could be used to label further samples, in order to train a larger and better model.

Rare Objects. Even if labelling effort would not be a limiting factor, and semi-automatic labelling would work, some objects occur only once or twice per tunnel, so that the training data set is in principle limited.

Object Variability. The types and appearance of objects varies between tunnels from different operators or due to specific local regulations. Thus processing a new tunnel may require adding new (variants) of objects.

Monitoring Quality. It is important that the quality of the segmentation model can be constantly monitored as new classes or training samples are added, in order to avoid performance drops on past data.

Human in the Loop. The process will always require a human operator to be involved. It is thus important that model can provide explanations that help understanding problems and guide the need for additional training data.

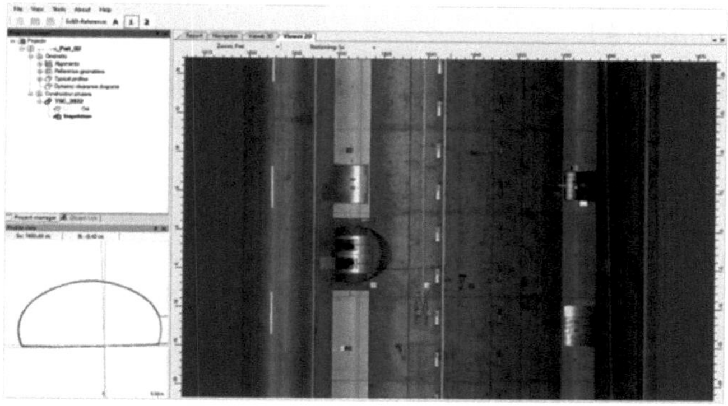

Fig. 1. Screenshot of the software Dibit8 showing 2D orthophoto plot of a street tunnel. The RGB photo data is used for annotating and labelling of AI training data.

In this work, we describe a workflow of object segmentation, including an approach based on few-shot object detection to handle cases with few training samples. In particular, we provide the following contributions:

- We extend the DE-ViT few-shot object detector with a DINOv3 backbone, CD-ViTO-style domain adaptation using learnable instance features, and hierarchical classification to reduce penalties for confusion between related classes. While DE-ViT provides only coarse object masks, these can guide subsequent tunnel object segmentation, though 3D segmentation is beyond this work.
- We leverage explainable visual analysis to diagnose the few-shot object detector, applying patch-level similarity maps and 2D embedding visualizations to identify visual cues supporting detections and analyze misclassifications within complex tunnel infrastructure scans.
- We introduce a patch overlap check based on the explaining visualisations to ensure semantic purity in class prototypes.

The remainder of this paper is organized as follows: Sect. 2 reviews related work, Sect. 3 details the detection framework and hierarchical classification, Sect. 4 describes XAI-based visualizations and resulting optimizations, Sect. 5 presents experimental results and Sect. 6 concludes the paper.

2 Related Work

2.1 Few-Shot Object Detection

Building upon the power of foundational models, recent work in few-shot object detection has explored leveraging pretrained backbones to address the challenge of learning novel categories with only a limited number of examples [13].

2.2 Explainable AI

Despite the impressive performance of complex deep learning models like DE-ViT, their inherent opacity, particularly when built upon large foundation models, poses challenges for understanding their decision-making processes [2]. This has driven significant attention to the field of explainable AI [2] mostly for the direct explanation of the foundation model features per se. For instance, DINOv2 patch embeddings have been used to highlight key features for explanation in re-identification tasks [4], and well known frameworks like Gradient-weighted Class Activation Mapping (GradCAM) [10] or SHapley Additive exPlanations (SHAP) [3] have been applied to models using DINOv2 for tasks like skin disease classification to pinpoint critical image regions [5] in medical applications. Visualisation approaches and related improvements for FSOD with DINOv2 have been proposed in [8], which we use as the basis of this work.

3 Proposed Approach

3.1 DINOv3-Enhanced DE-ViT Framework

We build on the DE-ViT architecture [13] and replace its DINOv2 backbone with the more robust DINOv3 [11], which provides improved training stability and higher-quality patch-level features due to large-scale pretraining.

Adapting DE-ViT to DINOv3 requires retraining the projection and detection heads on COCO and LVIS base classes, while keeping the DINOv3 backbone and the Region Proposal Network frozen. This ensures alignment with the new feature space while preserving the robustness of the pretrained representations, as detailed in [7].

3.2 Cross-Domain Adaptation with CD-VITO

For the rail tunnel application, we also investigated CD-ViTO (Cross-Domain Visual Transformer for Objects) [1], which is an extension of the DE-ViT architecture. It utilises a specialised, fine-tuning-based adaptation pipeline to overcome significant domain gaps. As this application involves substantial variation in the number of training samples per class, the instance reweighting module is not used. Instead, the model uses class-averaged mean prototypes to ensure a stable semantic representation, regardless of sample size. CD-ViTO focuses on learnable instance features, where initial fixed features are converted into learnable parameters. During fine-tuning, these parameters are optimised to align with target semantic concepts, increasing inter-class variance and ensuring the resulting mean prototypes are highly discriminative of both foreground objects and background instances. To further enhance robustness against environmental shifts, the domain prompter module synthesises virtual domain perturbations solely during finetuning stages. This mechanism ensures that the model learns to ignore varying image styles and focus on core semantic features, making the final prototypes highly resilient to the diverse lighting and textural conditions found in real-world target domains.

3.3 Modification of DE-ViT for Hierarchical Object Classification

Many tunnel objects can be assigned to a higher-level category, such as control box, box, or sign. We adapted the DE-ViT object detector to use this information for object detection. In the DE-ViT method, the confidence score is lower if an image region can be assigned to several object categories with similarly high confidence rather than to just one category. If a parent category has been specified, this check on the uniqueness of the assignment is suppressed for all classes sharing the same parent category.

4 Visual Explanations and Optimization

To analyze the performance of the few-shot detector and the quality of the training samples in the tunnel monitoring use case, we utilize visual explanations based on the DINO feature space. These methods allow for both localized patch-level analysis and a global view of how object categories are represented in the embedding space.

4.1 Patch-Level Visualization

In the DE-ViT framework, input images are divided into a grid of image patches (16×16 pixels) for tokenization. The feature vectors derived from these patches serve as the fundamental building blocks for detection: they are averaged within masked object regions to construct class prototypes and compared against query image patches via dot products to generate similarity scores for classification and localization. The implemented patch-level visualization method computes the similarity score between the feature vector of each image patch and the prototypes of the trained object classes. This is achieved by calculating the dot product between these vectors, highlighting which specific regions of an image contribute most strongly to a detection. For the visualization, the maximum similarity score from all prototypes (see Fig. 2) or from a specific class can be displayed.

Patch Overlap Check. During the analysis of training samples for tunnel objects, we found that very small object areas can be problematic. In the original DE-ViT implementation, an image patch is considered part of an object even if the overlap is as small as a single pixel. For small objects, this results in patches that contain a huge part of background area, leading to noisy prototypical feature vectors that include background information. To counteract this and ensure a more accurate semantic representation, we implemented a patch overlap check. In this optimized approach, we use only those patches for training that have an area overlap of at least 50% with the annotated object region (see Fig. 3). This filtering step provides substantial performance improvements (Sect. 5).

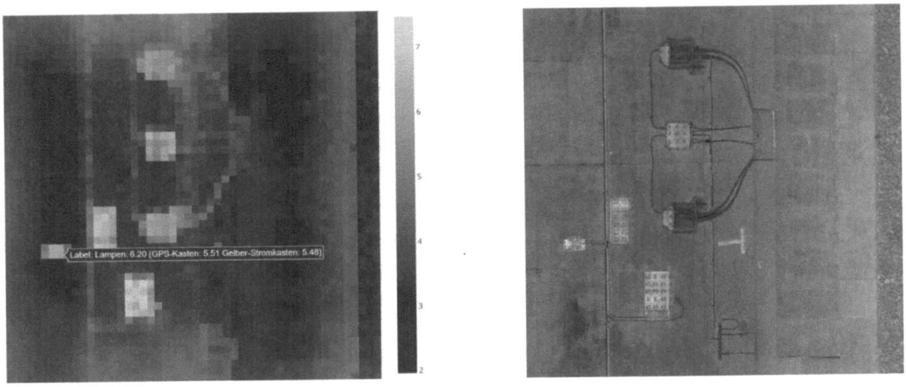

(a) (b)

Fig. 2. Patch-level visualization for the DE-ViT detector: (a) highest class similarity score per patch and showing top-3 class scores at the cursor position and (b) class index of the top score.

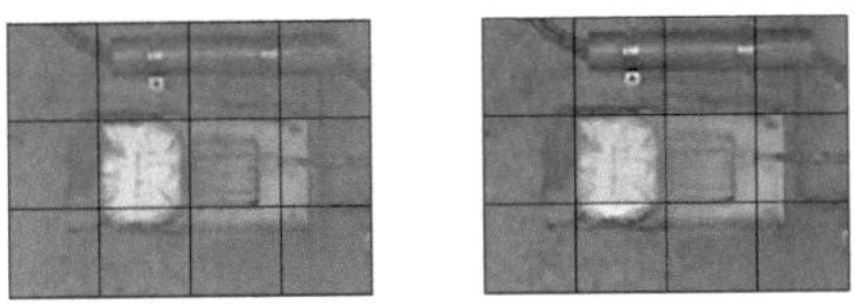

(a) (b)

Fig. 3. Comparison of two patch selection strategies for calculating the prototype feature vector: (a) selecting all patches overlapping an object by $\geq 1px$ (original implementation in DE-ViT), and (b) selecting only those with $\geq 50\%$ overlap.

Analysis of Misclassifications via Category-Specific Visualizations. In addition to displaying the maximum similarity score across all categories in the patch-level visualization, our framework allows for the visualization of individual similarity patch scores for a specific category. This detailed view is instrumental for the analysis of misclassifications, as it reveals whether an image region intended for one object class is being erroneously "attracted" to the prototype of another class. A clear instance of this was observed when analyzing the poor detection results for the "Umleitrolle" (deflection pulley) category using the standard DE-ViT detector, which achieved low performance (see Sect. 5). As shown in Fig. 4 in the upper row, the visualization of specific similarity scores reveals that the object regions for "Umleitrolle"—or significant portions of them—yielded high similarity scores for the "Erdung groß" (large grounding) category. This indicates that in the baseline DINO feature space, these two categories share overlapping features that the standard prototype-based projection cannot easily distinguish. CD-ViTO detector employs a specialized adaptation pipeline designed to increase inter-class variance. As demonstrated in Fig. 4, the interclass difference between the prototypes for "Umleitrolle" and "Erdung groß" is significantly higher when using CD-ViTO. This enhanced separation in the embedding space resulted in a substantial performance gain (see Sect. 5).

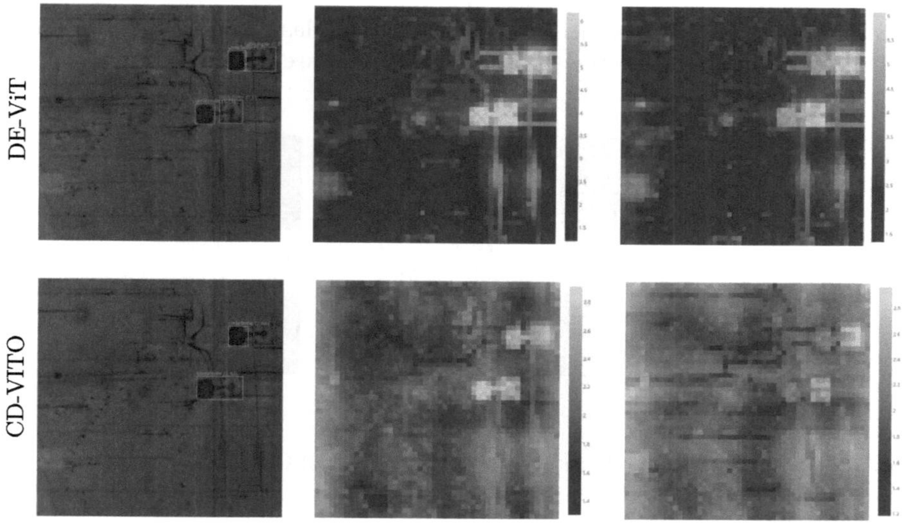

(a) Detection results (b) Similarity score 'Umleitrolle' (c) Similarity score 'Erdung groß'

Fig. 4. Visualisation of the similarity scores for the categories "Umleitrolle" and "Erdung groß" shows that the interclass difference between these categories is higher for CD-VITO than for DE-ViT, resulting in more reliable detection.

4.2 2D Embedding Visualization for Sample Analysis

To evaluate the relationships between different training samples and categories, we employ 2D embedding visualizations. By reducing the high-dimensional feature vectors to two dimensions using techniques like PCA or UMAP, we can inspect the distribution of samples and prototypes in an interactive environment using FiftyOne [6]. This visualization is a powerful tool for diagnosing model weaknesses:

Prototype Analysis: The 2D visualization can be used to display all class prototypes simultaneously (see Fig. 5). This allows us to identify classes with very similar feature vectors, which serves as a clear indicator that more or different training samples are required to distinguish between them and avoid misdetections.

Sample Refinement: In our tunnel monitoring dataset, we observed that the "Lampen" (lamps) category formed several distinct clusters. Using FiftyOne to inspect these clusters, we identified two groups consisting of training examples where the object was only partially visible at the left or right image borders (see Fig. 6(a) and (b)). By rejecting these non-representative samples, we were able to compute a more representative prototype for the class, leading to more stable and accurate detections in the tunnel environment (see Sect. 5).

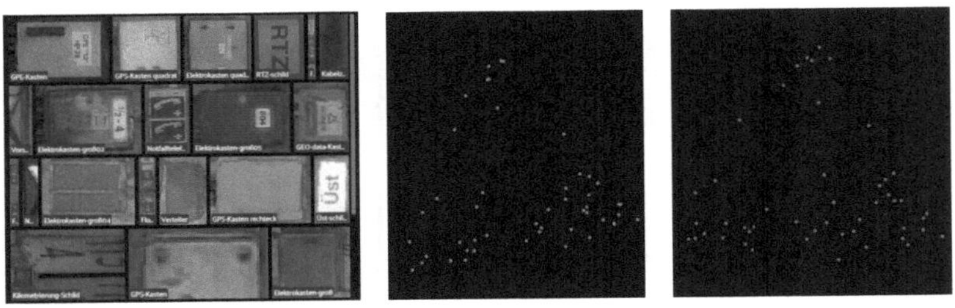

(a) (b) (c)

Fig. 5. 2D visualization of all prototypes ((b) DE-ViT, (c) CD-ViTO) and corresponding example image regions (a), created using FiftyOne.

In our previous research [8], we used the 2D representation of feature vectors from training samples for automatic cluster detection in multi-view representations. Although tunnel objects generally have fewer distinct views than general datasets, we can apply these techniques to identify outliers in the tunnel dataset as we have done manually for the "Lampen" category.

5 Experimental Results

5.1 Dataset

We use a dataset of 621 images (36,180 × 4,965 pixels) acquired with Dibit tunnel scanners, which are registered to a 3D point cloud of the tunnel. A 2D orthophoto derived from projections of true color values onto a reference tunnel profile represent a valuable data source to be used in an image-based object or damage segmentation approach. Such 2D orthophoto raster data sets contain radiometric information in three channels. The images were resized to a uniform height of 1,280 pixels, resulting in widths of approx. 9,327 pixels. To achieve square input images, tiles of 1,280 × 1,280 pixels were then sampled evenly from these resized images with minimum overlap. When splitting into training and validation datasets, care was taken that an entire input image with all the overlapping tiles was either in the training (80%) or the validation set (20%). CD-ViTO requires a final fine-tuning step when adding a new class. Using two NVIDIA RTX A6000, 500 training iterations for the rail tunnel dataset take about 18 min.

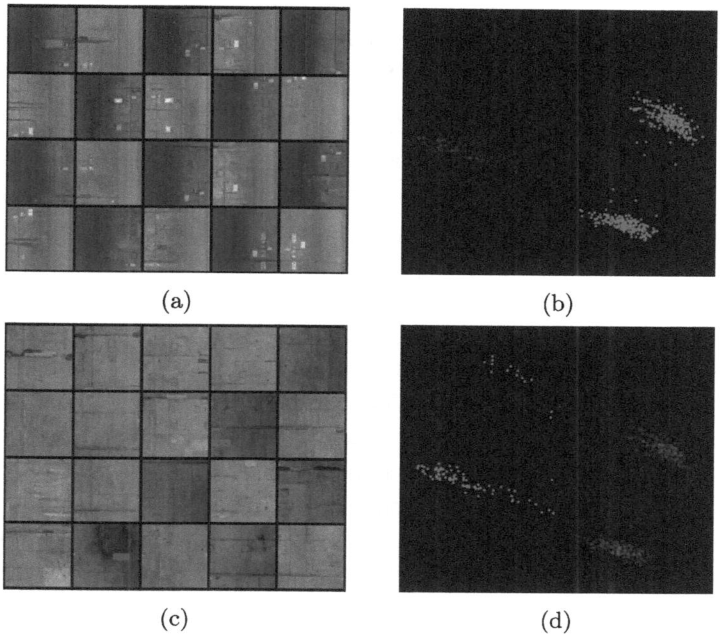

Fig. 6. Examples in the 2D embedding visualization (highlighted points on the right) and view of the related masked image regions (left) of class "Lampen".

5.2 Evaluation

We evaluate the proposed few-shot object detection approach on this rail tunnel dataset using DE-ViT and its domain-adaptive extension, CD-ViTO. For DE-ViT, we incorporate super-category information and improvements derived from visual analysis, including a patch overlap check for prototype computation and targeted sample refinement for the "Lampen" class. For CD-ViTO, we assess the impact of patch overlap and optional super-category integration. Results (Table 1) show that CD-ViTO significantly outperforms DE-ViT, with a substantial increase in mAP across all classes. This improvement is attributed to domain-adaptive fine-tuning, which transforms fixed features into discriminative, learnable representations. Moreover, enforcing a minimum 50% patch overlap reduces background noise and improves prototype quality. Additional sample filtering for "Lampen", increasing its mAP from 26.39% to 28.05%. Several categories resulted in 0% mAP due to extreme data scarcity ($n = 1$) (in the single validation image from "Antenne" and "Elektrokasten Rechteck" only a small part is visible) combined with high feature similarity to other classes (e.g. "Elektrokasten groß", "Elektrokasten-mittel02", and "Elektrokasten groß03").

Table 1. Final comparison of mAP results for DE-ViT and CD-VITO models. Tr = Training Samples, Val = Validation Samples, OC = Overlap Check, SC = Super Category, SR = Sample Refinement for category "Lampen"

Class	Tr	Val	DE-ViT				CD-VITO	
			Base	OC	OC+SC	OC+SC+SR	OC	OC+SC
Fluchtwegschilder	75	23	66.70	72.56	71.68	72.66	72.00	72.10
Elektrokasten	237	49	41.10	44.81	41.10	44.68	80.30	78.60
Lampen	743	165	0.10	25.49	26.39	28.05	36.40	36.40
Fluchtwegschilder-Klein	369	86	57.10	71.29	71.73	71.28	75.00	75.00
Kilometrierung-Schild	82	26	52.40	52.33	54.80	52.40	67.30	68.30
Gelber-Stromkasten	11	12	20.20	70.98	74.69	70.98	77.60	77.70
Bodenschacht	102	24	85.00	87.37	86.62	87.55	88.50	88.10
Elektrokasten-mittel	1	2	45.50	45.45	45.45	45.45	50.50	50.50
Elektrokasten-groß	1	3	0.00	0.00	0.00	0.00	0.00	0.00
Verteiler	1	1	0.00	0.00	0.00	0.00	0.00	0.00
Fluchtwegentfernung	88	23	64.70	75.22	73.24	75.33	84.70	83.00
Umleitrolle	28	9	6.30	7.72	4.78	7.36	65.70	65.70
Gegengewichtsanlage	60	16	8.40	13.10	13.67	12.87	66.20	61.90
Vorsignalschild	10	4	10.30	24.06	10.30	24.06	55.30	55.30
Erdung-groß	215	82	33.70	33.45	31.42	37.39	75.40	76.20
Erdung-klein	298	78	0.20	6.76	4.17	6.22	52.60	52.90
Signalstation	8	4	0.00	2.18	2.54	2.18	31.60	27.00
ANT	9	4	87.20	86.39	86.39	86.39	90.70	86.10
Notfalltelefon-Schild	12	2	29.60	60.13	60.13	60.13	46.70	46.70
Fluchtwegschild-groß	39	17	37.60	59.04	54.98	59.39	49.90	45.50
Erdung-klein02	250	142	0.00	3.56	2.45	3.52	23.30	26.00
Kabelzug Entlastung	30	2	0.00	0.00	0.00	0.00	5.10	20.20
Aufhängung	18	4	0.00	15.45	0.64	15.45	24.50	44.10
Elektrokasten quadrat	46	5	29.90	17.34	23.93	23.81	89.10	91.70
Elektrokasten Rechteck	4	1	0.00	0.00	0.00	0.00	0.00	0.00
Elektrokasten F	14	2	55.10	62.52	62.52	62.52	92.60	92.60
Elektrokasten Signalstation	2	2	53.40	56.76	71.85	56.76	92.60	92.60
GPS-Kasten quadrat	3	3	36.60	20.20	20.20	20.20	27.00	27.00
GPS-Station	8	2	30.30	30.30	30.30	30.30	47.50	47.10
Elektrokasten-GPS	2	2	24.60	55.05	50.00	55.05	80.00	80.00
GEO-data-Kasten	1	5	0.00	8.01	22.82	8.08	58.90	58.90
Antenne	2	1	0.00	0.00	0.00	0.00	0.00	0.00
GPS-Kasten rechteck	4	1	23.40	35.00	23.33	35.00	45.00	45.00
GPS-Kasten	4	4	0.00	0.45	1.20	0.44	12.20	24.50
Elektrokasten-groß04	1	1	0.00	0.00	0.00	0.00	0.00	0.00
Elektrokasten-groß05	3	2	90.00	92.52	92.52	92.00	92.60	92.60
Elektrokasten-mittel02	1	1	0.00	0.00	0.00	0.00	0.00	0.00
mAP	–	–	26.74	33.39	32.86	33.72	50.18	51.06

6 Conclusion and Outlook

This study demonstrates that few-shot object detection (FSOD) using the DE-ViT framework and its CD-ViTO extension effectively addresses the challenges of data scarcity and rare objects in rail tunnel monitoring. Visual explanations proved essential for bridging the gap between foundation models and domain-specific tasks: the implementation of a patch overlap check ensured semantic purity by filtering noisy background data from prototypes. Furthermore, 2D embedding visualizations facilitated the identification and rejection of non-representative outliers leading to more stable and accurate detections. Domain-adaptive fine-tuning in CD-ViTO significantly enhances performance, and that insights from XAI remain vital for model robustness and interpretability in human-in-the-loop workflows. Future research will focus on using adversarial examples, created with guidance from data gathered for the visualisations, in order to provide support for improving the sample set.

Acknowledgments. This work has received funding from the Austrian Federal Ministry of Innovation, Mobility and Infrastructure (BMIMI) in the project "FAIR-AI" funded under the AI AUSTRIA Initiative and the project "plAIdata" funded under the "Digital Technologies" program.

Disclosure of Interests. The authors have no competing interests to declare that are relevant to the content of this article.

References

1. Fu, Y., et al.: Cross-domain few-shot object detection via enhanced open-set object detector. In: Leonardis, A., Ricci, E., Roth, S., Russakovsky, O., Sattler, T., Varol, G. (eds.) Computer Vision - ECCV 2024, pp. 247–264. Springer, Cham (2025)
2. Kazmierczak, R., Berthier, E., Frehse, G., Franchi, G.: Explainability for vision foundation models: a survey (2025). https://doi.org/10.48550/arXiv.2501.12203
3. Lundberg, S.M., Lee, S.I.: A unified approach to interpreting model predictions. In: Advances in Neural Information Processing Systems, vol. 30 (2017)
4. Maddigan, P., Ehrhardt, O., Lensen, A., Shaw, R.C.: Re-identification of individual Kākā: an explainable DINO-based model. In: 2024 39th International Conference on Image and Vision Computing New Zealand (IVCNZ), pp. 1–6. IEEE (2024)
5. Mohan, J., Sivasubramanian, A., Ravi, V.: Enhancing skin disease classification leveraging transformer-based deep learning architectures and explainable AI. Comput. Biol. Med. **190**, 110007 (2025)
6. Moore, B.E., Corso, J.J.: Fiftyone. GitHub (2020). https://github.com/voxel51/fiftyone
7. Neuschmied, H., Bailer, W., Winter, M.: DINOv3 for DE-ViT: boosting few-shot object detection with scalable and robust visual features. In: International Conference on Computer Vision Theory and Applications (2026)
8. Neuschmied, H., Winter, M., Bailer, W.: Improving few-shot object detection using visual explanations of DINOv2 features. In: International Conference on Multimedia Modeling, pp. 59–71. Springer (2026)

9. Ren, D., Li, J., Wu, Z., Guo, J., Wei, M., Guo, Y.: MFFNet: multimodal feature fusion network for point cloud semantic segmentation. Vis. Comput. **40**(8), 5155–5167 (2024)
10. Selvaraju, R.R., Cogswell, M., Das, A., Vedantam, R., Parikh, D., Batra, D.: Grad-CAM: visual explanations from deep networks via gradient-based localization. In: Proceedings of the IEEE International Conference on Computer Vision, pp. 618–626 (2017)
11. Siméoni, O., et al.: DINOv3 (2025). https://arxiv.org/abs/2508.10104
12. Yue, H., Wang, Q., Zhang, M., Xue, Y., Lu, L.: 2D–3D fusion approach for improved point cloud segmentation. Autom. Constr. **177**, 106336 (2025)
13. Zhang, X., Liu, Y., Wang, Y., Boularias, A.: Detect everything with few examples (arXiv:2309.12969) (2024). https://doi.org/10.48550/arXiv.2309.12969

NExEM: Model-Agnostic Natural Language Explanation of Entity Matching

Hironori Mizuguchi[1,2(✉)] and Hiroyuki Kitagawa[1]

[1] University of Tsukuba, Tsukuba, Japan
s2230158@u.tsukuba.ac.jp, hmizuguc@lycorp.co.jp, kitagawa@cs.tsukuba.ac.jp
[2] LINE Yahoo Corporation, Tokyo, Japan

Abstract. Explainability is increasingly important in Entity Matching, where models decide whether two records refer to the same real-world entity. Existing post-hoc methods such as LIME and SHAP provide token-level attribution scores, but these numeric outputs are difficult for non-experts to interpret and do not reflect the comparative reasoning central to entity matching. Natural language explanation methods generate readable rationales, yet are typically tied to specific models. We propose NExEM, a post-hoc, model-agnostic framework that converts attribution scores from any XAI method into fluent explanations using a large language model. NExEM uses record-pair-aware prompting to structure explanations around shared attributes, differences, and missing information. Experiments on public datasets with two matching models, including user studies and LLM-based evaluation, show that NExEM produces significantly more understandable explanations than a baseline prompt.

Keywords: Explainable AI · Entity Matching · Machine Learning

1 Introduction

Data integration is a fundamental challenge in data management, encompassing tasks such as schema matching, data deduplication, and entity matching. Among these, entity matching—determining whether two records from different data sources refer to the same real-world entity—plays a central role in applications such as knowledge base construction, product catalogue integration, and customer data consolidation.

As machine learning models have become the dominant approach for entity matching, concerns about their interpretability have grown. In many practical settings, it is not sufficient for a system to output only a binary match/non-match decision; users and domain experts also require an explanation of *why* two records were judged to match or not. Such explanations are important for debugging model errors, auditing decisions, and building trust in automated data integration pipelines.

© The Author(s), under exclusive license to Springer Nature Switzerland AG 2027
M. Geiß et al. (Eds.): DEXA 2026 Workshops, CCIS 3078, pp. 27–38, 2027.
https://doi.org/10.1007/978-3-032-34202-7_3

Explainable AI (XAI) methods such as LIME [16] and SHAP [13] have been widely used to address this need. These model-agnostic methods provide local explanations in the form of token-level attribution scores. Several studies have adapted them to entity matching [1,2,4,6]. However, attribution-based explanations have a key limitation: they present tokens with numerical scores, which are difficult for non-experts to interpret and fail to capture the comparative reasoning underlying entity matching decisions.

A complementary line of research explores natural language explanations (NLEs), in which a model generates a free-text rationale alongside its prediction [5]. Although more accessible to users, existing NLE methods for entity matching are tightly coupled with specific matching models and cannot be applied post-hoc to arbitrary classifiers.

To address this gap, we propose Natural-language EXplanation for Entity Matching (NExEM), a framework that combines the model-agnosticism of attribution-based XAI with the accessibility of natural language explanations. NExEM takes token-level attribution scores produced by existing XAI methods—such as LIME, SHAP, Landmark, or LEMON—and converts them into fluent explanations using an LLM. Its key contribution is a structured prompt with *record-pair-aware guidelines*, which instructs the LLM to organize explanations around matching points, differing points, and missing information supported by the attribution scores.

The contributions of this paper are as follows:

- We identify the gap between attribution-based XAI and natural language explanations for entity matching, and propose a post-hoc framework that bridges the two.
- We design a structured prompt with record-pair-aware guidelines that grounds explanations in the comparative structure of entity matching decisions.
- We evaluate the framework on benchmark datasets through human evaluation and an LLM-as-a-judge protocol, showing that the proposed guidelines produce significantly more understandable explanations than a baseline prompt.

The remainder of this paper is organized as follows. Section 2 reviews related work. Section 3 formalizes the problem. Section 4 describes the framework. Section 5 presents the experiments. Section 6 concludes the paper.

2 Related Work

2.1 Entity Matching

Entity matching has been approached through a variety of methods, including rule-based and machine-learning-based techniques.

Deep-learning-based approaches, in particular, have attracted significant attention from researchers in recent years [3]. DeepER [8] represents each record as a vector using an LSTM architecture with GloVe word embeddings and uses the similarity between the two vectors for entity matching. DeepMatcher [14] assumes a known correspondence between attributes in the two datasets being compared: an attribute-embedding module embeds the values of each attribute, an attribute-similarity representation module generates a similarity vector from these embeddings, and a classification module predicts whether the pair constitutes a match. DITTO [12] is a transformer-based approach that serialises each record into a text sequence comprising attribute names and values, then feeds the pair into a pretrained BERT model to determine whether the records match. DITTO also provides extensions such as knowledge injection via named entity recognition (NER) tags.

More recent work has examined the use of large language models for entity matching. Peeters et al. [15] investigated zero-shot and few-shot prompting with generative LLMs, and showed that GPT-4 can generate structured explanations and identify error causes from mispredictions. Table-GPT [11] introduces a table-tuning approach where GPT models are fine-tuned on diverse table tasks including entity matching, demonstrating the benefits of tailoring LLMs for tabular data.

2.2 Explainable AI for Entity Matching

Explainable AI (XAI) has been extensively studied in recent years [7,9]. Among general-purpose XAI methods, LIME [16] and SHAP [13] are notable in that both are *model-agnostic* approaches that provide local explanations. LIME produces attribution scores by approximating a complex model with a linear surrogate fitted to perturbed samples. SHAP is inspired by Shapley values from cooperative game theory and provides a rigorous mathematical framework for computing feature attributions.

Mojito [6] uses LIME to provide attribute-level explanations for entity matching, specifically addressing cases where a few attributes contribute to a non-match decision by augmenting records and observing changes in match scores. Landmark [1] is a word-level explanation method that applies LIME with a two-sided perturbation strategy, fixing each record in turn and perturbing the other, producing a dual explanation. LEMON [2] adopts a similar dual-explanation strategy but additionally augments tokens from the unfixed record into the fixed record to reveal how a non-match might become a match. CREW [4] produces token cluster-level explanations by building a signed, weighted knowledge graph over tokens and applying correlation clustering to partition them into coherent groups ranked by their influence.

2.3 Natural Language Explanations for Entity Matching

Natural language explanations (NLEs) have been studied as a means of making AI decisions more transparent and accessible [5]. Unlike feature attribution methods, NLEs express the reasoning behind a prediction in free-form text, making them more interpretable to non-expert audiences.

Wadhwa et al. [18] explored leveraging natural language explanations to improve the generalizability of EM models. They noted that directly applying large language models to entity matching is often prohibitively expensive. To address this, they reformulated entity matching as a conditional text generation problem, where the model generates not only the label (match/unmatch) but also a supporting rationale, i.e., a natural-language explanation justifying the decision. They further distilled these rationales produced by large language models into smaller sequence-to-sequence models, which can be trained more efficiently. Their study demonstrates that incorporating instance-specific natural language explanations can substantially improve out-of-distribution performance.

While these approaches demonstrate the value of LLM and natural language explanations for entity matching, they generate explanations as an intrinsic component of the matching model itself. Consequently, they are not model-agnostic and cannot be applied post-hoc to arbitrary classifiers. In contrast, NExEM decouples the explanation generation step from the matching model, applying an existing XAI method to any classifier and subsequently translating the resulting token attributions into natural language via an LLM.

3 Problem Definition

Entity Matching. Each entity is a real-world object typically represented as a structured record of attribute name and attribute value pairs, where the values may be textual, numerical, categorical, or of other types. In this study, we assume that all attribute values are converted into strings.

Let $\mathcal{D}_1$ and $\mathcal{D}_2$ be two datasets. Given a record pair (r_1, r_2) where $r_1 \in \mathcal{D}_1$ and $r_2 \in \mathcal{D}_2$, the entity matching task is to predict a binary label $y \in \{match, non\text{-}match\}$ indicating whether r_1 and r_2 refer to the same real-world entity. We assume access to a trained matching model f that returns a prediction score $f(r_1, r_2) \in [-1, 1]$, where 1 and -1 represent the strongest *match* and *non-match*, respectively.

Token-Level Attribution. We assume access to an existing XAI method—such as LIME, SHAP, or any of their variants tailored to entity matching—that produces a token-level attribution score $\phi_t \in \mathbb{R}$ for each token t appearing in attribute values of r_1 or r_2. Here, a *token* denotes the smallest unit to which an XAI method assigns an attribution score; in our setting, each token corresponds to an individual word in an attribute value. A positive score of ϕ_t indicates evidence in favour of a match decision; a negative score indicates evidence in favour of a non-match decision. The magnitude of ϕ_t reflects the strength of that evidence. Our framework treats the XAI method as a black box and imposes no constraints on how these scores are computed.

Explanation. Given a record pair (r_1, r_2), a prediction $\hat{y}$, the prediction score $f(r_1, r_2)$, and token-level attribution scores $\{\phi_t\}$, the goal of our framework is to generate a natural language explanation e that satisfies the following requirements:

1. **Understandability**: e should be comprehensible to users with no background in machine learning.
2. **Record-awareness**: e should articulate the relationship between the two records explicitly, covering matching points, differing points, and missing information where supported by the scores.
3. **Fidelity**: e should accurately reflect the attribution scores $\{\phi_t\}$.
4. **Model-agnosticism**: the framework should be applicable to any matching model f and any XAI method without modification.

4 Proposed Method: NExEM

4.1 Overview

NExEM framework takes as input the token-level attribution scores produced by an arbitrary XAI method applied to an entity matching model, and generates a natural language explanation of the matching decision via an LLM. The pipeline consists of a single natural language generation stage driven by a carefully designed prompt, and is applicable to any combination of matching model and XAI method.

4.2 Prompt Design

The prompt provided to the LLM consists of three components: a *task description*, an *interpretation guide*, and *record-pair-aware guidelines*.

Task Description. The task description specifies the role of the LLM and the structure of the input. The LLM is instructed to act as an assistant explaining an entity matching decision to a non-expert user. The input includes the item category, the two records r_1 and r_2, the matching decision $\hat{y}$ and its score $f(r_1, r_2)$, and the token-level attribution scores $\{\phi_t\}$.

Interpretation Guide. The interpretation guide maps attribution score ranges to qualitative evidence labels, enabling the LLM to ground its explanation in the numerical scores without requiring the user to interpret raw numbers. Specifically, scores of 0.5 or above indicate strong evidence for a match; scores between 0.1 and 0.5 indicate moderate evidence; scores between 0 and 0.1 indicate weak evidence; and the symmetric ranges apply for non-match evidence.

Record-Pair-Aware Guidelines. The central contribution of our prompt design is a set of record-pair-aware guidelines that explicitly direct the LLM to organize the explanation around the comparative relationship between the two records. The guidelines instruct the LLM to cover three aspects where supported by the attribution scores: (a) *matching points*, where similar or identical information appears in both records; (b) *differing points*, where the two records describe information differently or inconsistently; and (c) *missing information*, where an important detail appears in only one record. Aspects not clearly supported by the scores may be omitted.

The full prompt template is shown in Fig. 1.

```
You are an assistant that explains an AI system's Entity Matching
  decision to non-expert users with no background in machine
    learning.

Input consists of:
- Item category (e.g., product, company)
- Two items (Item 1 and Item 2)
- The AI's decision (Same or Different) and its score
- Token-level contribution scores

Interpretation of token-level contribution scores:
- A score of 0.5 or higher: strong evidence supporting the Same
  decision.
- A score between 0.1 and 0.5: moderate evidence supporting the
  Same decision.
- A score between 0 and 0.1: weak evidence supporting the Same
  decision.
- A score between 0 and -0.1: weak evidence supporting the
  Different decision.
- A score between -0.1 and -0.5: moderate evidence supporting the
  Different decision.
- A score of -0.5 or lower: strong evidence supporting the
  Different decision.

Task:
Generate a natural language explanation of why the AI produced
this decision
so that a general audience can easily understand it,
based only on the given token-level contribution scores.

Output constraints:
- The total length should be around 60 words (approximately
  50-70 words).

[Guidelines:
- Organize the explanation using bullet points.
- Cover the following aspects when they are clearly supported by
  the token-level contribution scores:
  (a) matching points, where similar or identical information
      appears in both Item 1 and Item 2;
  (b) differing points, where the two items describe information
      differently or inconsistently;
  (c) missing information, where an important detail appears in
      only one item.
- If any aspect is not clearly supported by the scores, you may
  omit that bullet point.
- Focus on helping the reader understand how the two items relate
  to each other.]

Restrictions:
- Do NOT speculate beyond the given scores.
- Use English only.
```

Fig. 1. System prompt template used for explanation generation. The guidelines block (marked in brackets) is included in NExEM and omitted in the baseline.

4.3 Effect of Record-Pair-Aware Guidelines

To investigate the contribution of the record-pair-aware guidelines, we compare two prompt variants: the *full prompt* including all three components described above (NExEM), and a *baseline prompt* from which the guidelines component is removed (Baseline). The baseline prompt thus instructs the LLM to generate a natural language explanation without explicit guidance on how to structure the comparison between the two records. The effect of this design choice is evaluated in Sect. 5.

5 Experiments

5.1 Experimental Setup

Datasets. We evaluate our framework on 12 datasets from DeepMatcher [14], which includes three dataset types: Structured, Dirty, and Textual. "Structured" denotes relatively clean benchmark datasets with well-aligned attribute values, "Dirty" denotes noisier versions in which some words within attribute values are deleted or replaced, and "Textual" denotes datasets with longer free-text fields. Table 1 lists the datasets used.

Entity Matching Models. We use two entity matching models as the underlying classifiers. The first is DITTO [12], The second model is provided by py_entitymatching [17], an implementation of Magellan [10], which uses gradient-boosted decision trees (GBDT). These two models represent contrasting paradigms—deep learning versus traditional feature engineering—and their inclusion demonstrates that our explanation framework is applicable to diverse underlying classifiers.

Table 1 reports the precision, recall, and F1 scores of both models across all datasets.

Table 1. Datasets used in our experiments and entity matching model performance.

Type	Dataset	DITTO			py_entitymatching		
		Prec.	Rec.	F1	Prec.	Rec.	F1
Structured	Amazon-Google	0.778	0.664	0.717	0.457	0.690	0.550
	Beer	1.000	0.583	0.737	1.000	0.737	0.849
	DBLP-ACM	0.984	0.984	0.984	0.996	0.978	0.987
	DBLP-GoogleScholar	0.958	0.934	0.946	0.932	0.934	0.933
	Fodors-Zagats	0.955	1.000	0.977	1.000	1.000	1.000
	Walmart-Amazon	0.777	0.824	0.800	0.529	0.872	0.658
	iTunes-Amazon	0.963	0.839	0.897	0.963	0.839	0.897
Dirty	DBLP-ACM	0.982	0.971	0.977	0.932	0.892	0.912
	DBLP-GoogleScholar	0.965	0.932	0.948	0.812	0.850	0.830
	Walmart-Amazon	0.793	0.827	0.810	0.301	0.725	0.425
	iTunes-Amazon	0.963	0.813	0.881	0.519	0.609	0.560
Textual	Abt-Buy	0.850	0.818	0.833	0.393	0.794	0.526

Token-Level Attribution Method. For all experiments, token-level attribution scores are computed using LIME applied to each matching model. Each token in the concatenated record pair is treated as a feature, and attribution scores are computed independently for each record pair.

Explanation Conditions. We compare two prompt variants NExEM and Baseline, which differ only in the presence or absence of the record-pair-aware guidelines. In both conditions, the task description and interpretation guide remain identical, isolating the effect of the record-pair-aware guidelines. Attribution scores in both conditions are top-5-ranked features for each record.

We use gemini-2.5-flash as the LLM for explanation generation in all experiments. Explanations are generated each time for each record pair and prompt condition without reuse.

Figure 2 shows an example of a user prompt and the generated explanations for both NExEM and Baseline conditions.

5.2 Experiment 1: User Study

Setup. To assess the understandability of the generated explanations from a human perspective, we conducted a user study via Amazon Mechanical Turk. For each of the 12 datasets, we sampled 2 matched record pairs and 2 non-matched record pairs from the predictions of the DITTO model, yielding 48 record pairs in total. For each pair, we generated two explanations—NExEM and Baseline—using LIME-based attribution scores from DITTO. Annotators were presented with both explanations and asked to select which explanation was clearer and easier to understand. To ensure task comprehension, each question included a

(a)

```
-- Item category --
beer

-- Item 1 --
- Beer_Name: Bourbon Barrel Aged Horus Imperial Red
- Brew_Factory_Name: Elevator Brewery & Draught Haus
- Style: American Amber / Red Ale
- ABV: 12.00 %

-- Item 2 --
- Beer_Name: Lompoc Bourbon Barrel Aged Old Tavern Rat
- Brew_Factory_Name: Lompoc Brewing
- Style: Barley Wine
- ABV: 9.70 %

-- AI system's Entity Matching decision --
Different items (-0.96 - strong evidence)

-- Token-level contribution scores --
(Ranked by impact)

Item 1:
- Draught: -0.05292832682944003
- Elevator: -0.0494520793732402
- /: -0.041770254467486334
- Bourbon: 0.027514097004016642
- Barrel: 0.02491493451121246

Item 2:
- Brewing: -0.39747562002478953
- Wine: -0.15490768217606693
- Lompoc: -0.14034328885775363
- Barley: -0.10926217221725738
- Barrel: 0.020261441990006145
```

(b)

```
The AI determined these are different beer items due to several
key differences:

* **Differing points:** The AI found significant differences
  in the brewery names, particularly "Lompoc Brewing" in Item 2.

* The beer styles also differed, with "Barley Wine" in Item 2
  contributing to the "Different" decision.

* **Matching points:** The terms "Bourbon" and "Barrel"
  appeared in both beer names, offering a slight, weak
  indication of similarity.
```

(c)

```
The AI determined these beers are different. Words like "Brewing,"
"Wine," "Lompoc," and "Barley" from the second beer strongly
indicated they are not the same. Additionally, "Draught" and
"Elevator" from the first beer also suggested differences. While
"Bourbon" and "Barrel" showed slight similarities, the strong
differences outweighed them.
```

Fig. 2. Example of user prompt used and generated explanation. (a) user prompt, (b) explanation generated by NExEM, and (c) explanation generated by the baseline.

quality check asking annotators to confirm the model's decision. Each record pair was evaluated by 3 annotators.

Results. Table 2 summarises the results. Overall, annotators preferred NExEM in 137 out of 144 valid comparisons (95.1%), a proportion significantly greater than chance ($p = 1.03 \times 10^{-32}$, one-sided binomial test). The preference for NExEM was consistent across both matched pairs and non-matched pairs, demonstrating that record-pair-aware guidelines substantially improve the understandability of explanations regardless of the matching decision.

Table 2. User study results. Annotators selected which of two explanations was clearer. p-values are from a one-sided binomial test.

Group	Total	NExEM	Baseline	Ratio	p-value
Overall	144	137	7	0.951	1.03×10^{-32}
Match	72	71	1	0.986	1.54×10^{-20}
Non-match	72	66	6	0.916	3.62×10^{-14}

5.3 Experiment 2: LLM-as-a-Judge Evaluation

Setup. To complement the user study with a larger-scale evaluation, we employed an LLM-as-a-judge protocol. For each of the 12 datasets, we sampled 20 matched and 20 non-matched record pairs, yielding 480 record pairs per model. We applied our framework to both DITTO and `py_entitymatching`, generating explanations under both prompt conditions (NExEM and Baseline) using LIME-based attributions. An LLM judge was then presented with both explanations for each record pair and asked to select the clearer and more understandable explanation, following the same comparative format as the user study.

Results. Table 3 report the results for DITTO and `py_entitymatching`. For DITTO, NExEM was preferred in 416 out of 480 comparisons (86.7%), with a highly significant overall preference ($p = 1.46 \times 10^{-64}$, one-sided binomial test). The advantage held consistently across matched pairs and non-matched pairs. For `py_entitymatching`, NExEM was preferred in 417 out of 479 comparisons (87.1%), again with a highly significant overall preference ($p = 5.83 \times 10^{-66}$).

Taken together, the results across both models demonstrate that record-pair-aware guidelines consistently improve the understandability of generated explanations, regardless of the underlying matching model. This confirms the model-agnostic nature of our framework.

Table 3. LLM-as-a-judge results for DITTO and `py_entitymatching`.

Group	DITTO				`py_entitymatching`			
	NExEM	Base	Ratio	*p*-value	NExEM	Base	Ratio	*p*-value
Overall	416	64	0.867	1.46×10^{-64}	417	62	0.871	5.83×10^{-66}
Match	209	31	0.871	6.51×10^{-34}	199	40	0.833	7.44×10^{-27}
Non-match	207	33	0.863	2.72×10^{-32}	218	22	0.908	4.80×10^{-42}

Consistency Between Evaluations. The results of the user study and the LLM-as-a-judge evaluation are consistent in direction: both show a strong and statistically significant preference for explanations generated with record-pair-aware guidelines across all dataset types and matching decisions. The slightly lower preference ratio in the LLM-as-a-judge evaluation (86.7–87.1% vs. 95.1% in the user study) may reflect differences in evaluation criteria between human annotators and the LLM judge, but does not alter the overall conclusion.

5.4 Additional Experiments on Faithfulness

To assess whether the generated explanations faithfully reflect the underlying entity matching decisions and attribution features, we conduct two additional experiments. Both experiments are applied only to explanations generated by NExEM for DITTO.

Experiment 3: Decision Faithfulness

Setup. A natural language explanation should be consistent with the matching decision it is intended to explain. To verify this property, we evaluate whether the generated explanation alone is sufficient to recover the original matching decision. Specifically, we present each explanation to an LLM without providing the original records or the model's prediction score, and ask it to infer whether the decision was *match* or *non-match* based solely on the explanation text. We then compare the inferred decision with the actual decision produced by the entity matching model, and report accuracy for the Overall, Match, and Non-match groups.

If an explanation accurately conveys the reasoning behind a matching decision, an LLM should be able to recover the correct decision with high accuracy. Conversely, low accuracy would suggest that the explanation contains hallucinated or misleading content inconsistent with the actual decision.

Results. Table 4 reports the results. Overall, NExEM enabled correct recovery of the original matching decision from the explanation in 473 out of 480 cases (98.5% accuracy). The accuracy was similarly high for both matched pairs (98.3%) and non-matched pairs (99.2%), demonstrating that the generated explanations are highly faithful to the underlying matching decisions across all dataset types and decision categories.

Table 4. Decision faithfulness: accuracy of LLM-inferred decisions from explanations alone.

Group	Correct	Total	Accuracy
Overall	473	480	98.5%
Match	235	240	98.3%
Non-match	238	240	99.2%

Experiment 4: Attribution Faithfulness

Setup. Consistency with the final prediction does not necessarily imply faithfulness to the evidence used by the model. Because NExEM takes token-level attribution scores as input, its explanations should accurately reflect those scores. To evaluate this property, we present the LLM judge with the record pair, the top-5 token-level attribution scores, and the generated explanation. The judge then answers three questions for the tokens referenced in the explanation: 1) Whether all referenced tokens appear in the input records, 2) Whether all referenced tokens are included in the provided attribution scores, 3) Whether the explanation correctly reflects the sign of each score (i.e., supporting *match* or *non-match*).

The first two questions assess grounding in valid evidence, while the third evaluates correct interpretation of that evidence.

Results. Table 5 reports the results. Overall, 98.1% of referenced tokens appeared in the input records, indicating that hallucinated tokens were rare. Moreover, 92.5% were included in the attribution scores, showing that explanations were largely grounded in attributed evidence.

The accuracy for correctly reflecting attribution direction was slightly lower at 85.4%. Thus, explanations generally selected appropriate tokens, but did not always express whether they supported *match* or *non-match* correctly.

Performance was consistently higher for the Match group than for the Non-match group, suggesting that negative evidence is harder to interpret.

Qualitative inspection further revealed Match cases where a shared token received a negative attribution score. Some explanations described such tokens positively or ambiguously, reversing the intended meaning. Future work will investigate these cases further.

Table 5. Attribution faithfulness of generated explanations.

Group	Tokens in records	Tokens in attributions	Correct direction
Overall	471/480 (98.1%)	444/480 (92.5%)	410/480 (85.4%)
Match	239/240 (99.6%)	232/240 (96.7%)	215/240 (89.6%)
Non-match	232/240 (96.7%)	212/240 (88.3%)	195/240 (81.3%)

6 Conclusion

In this paper, we presented NExEM, a post-hoc and model-agnostic framework for generating natural language explanations of entity matching decisions. NExEM takes token-level attribution scores produced by existing XAI methods, such as LIME or SHAP, and converts them into fluent, human-readable explanations using a large language model. Its key contribution is a structured prompt with record-pair-aware guidelines that organize explanations around matching points, differing points, and missing information.

We evaluated the framework on benchmark datasets. The results showed that explanations generated with the proposed guidelines were significantly more understandable than those generated without such structure.

These findings suggest that attribution-based XAI can be effectively translated into natural language explanations through prompt design, without modifying the underlying matching model or XAI method. This makes the framework a practical post-hoc explanation layer for existing entity matching systems.

Limitations and Future Work. Although NExEM is model-agnostic, the current study focused mainly on LIME-based attribution scores. Evaluating other XAI methods such as SHAP and Landmark remains future work. Attribution-based explanations may inherit instability from the underlying XAI method. We also observed that verbalizing attribution direction is particularly challenging for non-match cases. In addition, evaluations relying on LLM judges may introduce evaluator bias. Future work includes investigating advanced prompting strategies and different LLMs, including smaller locally deployable models.

Acknowledgments. This work was supported by JSPS KAKENHI (JP23K28089, JP25K22810) and AMED (JP21zf0127005).

References

1. Baraldi, A., Del Buono, F., Paganelli, M., Guerra, F.: Landmark explanation: an explainer for entity matching models. In: Proceedings of CIKM 2021, pp. 4680–4684 (2021)
2. Barlaug, N.: Lemon: explainable entity matching. IEEE Trans. Knowl. Data Eng. **35**(8), 8171–8184 (2023)
3. Barlaug, N., Gulla, J.A.: Neural networks for entity matching: a survey. ACM Trans. Knowl. Discov. Data **15**(3) (2021)
4. Benassi, R., Guerra, F., Paganelli, M., Tiano, D.: Explaining entity matching with clusters of words. In: Proceedings of ICDE2024, pp. 2325–2337 (2024)
5. Camburu, O.M., Rocktäschel, T., Lukasiewicz, T., Blunsom, P.: e-SNLI: natural language inference with natural language explanations. In: NIPS'18, pp. 9560–9572 (2018)
6. Di Cicco, V., Firmani, D., Koudas, N., Merialdo, P., Srivastava, D.: Interpreting deep learning models for entity resolution: an experience report using LIME. In: Proceedings of aiDM '19, pp. 1–4 (2019)
7. Du, M., Liu, N., Hu, X.: Techniques for interpretable machine learning. Commun. ACM **63**(1), 68–77 (2019)
8. Ebraheem, M., Thirumuruganathan, S., Joty, S., Ouzzani, M., Tang, N.: Distributed representations of tuples for entity resolution. Proc. VLDB Endow. **11**(11), 1454–1467 (2018)
9. Hind, M.: Explaining explainable AI. XRDS **25**(3), 16–19 (2019)
10. Konda, P., et al.: Magellan: toward building entity matching management systems. Proc. VLDB Endow. **9**(12), 1197–1208 (2016)
11. Li, P., et al.: Table-GPT: table fine-tuned GPT for diverse table tasks. Proc. ACM Manag. Data **2**(3) (2024)
12. Li, Y., Li, J., Suhara, Y., Doan, A., Tan, W.C.: Deep entity matching with pre-trained language models. Proc. VLDB Endow. **14**(1), 50–60 (2020)
13. Lundberg, S.M., Lee, S.I.: A unified approach to interpreting model predictions. In: Proceedings of NIPS 2017. NIPS'17, pp. 4768–4777 (2017)
14. Mudgal, S., et al.: Deep learning for entity matching: a design space exploration. In: Proceedings of SIGMOD '18, pp. 19–34 (2018)
15. Peeters, R., Steiner, A., Bizer, C.: Entity matching using large language models (2024). https://arxiv.org/abs/2310.11244
16. Ribeiro, M.T., Singh, S., Guestrin, C.: "Why should I trust you?": Explaining the predictions of any classifier. In: Proceedings of KDD '16, pp. 1135–1144 (2016)
17. Team, U.M.: py_entitymatching 0.4.1. https://pypi.org/project/py-entitymatching/
18. Wadhwa, S., Krishnan, A., Wang, R., Wallace, B.C., Kong, L.: Learning from natural language explanations for generalizable entity matching. In: Proceedings of EMNLP '24, pp. 6114–6129 (2024)

Enhancing E-Commerce Product Data with AI at PBS Holding: Nearest Neighbor eClass Classification from Product Text Embeddings

Clara Schachner[1(✉)], Kyrylo Meliushko[1], Florian Kromp[1], Paria Huemer[2], Michael Schuch[2], Wolfgang Salm[2], and Raphael Pisoni[1]

[1] Software Competence Center Hagenberg, Hagenberg, Austria
{clara.schachner,kyrylo.meliushko,florian.kromp,raphael.pisoni}@scch.at
[2] PBS Holding, Wels, Austria
{paria.huemer,michael.schuch,wolfgang.salm}@pbs-holding.at

Abstract. Accurate e-commerce product data is essential for effective search, filtering, and product discovery, directly influencing customer satisfaction, sales performance, and operational efficiency. However, vendor provided descriptions and images are often incomplete or misleading, reducing data quality and affecting downstream processes. This study presents an eClass classification use case, developed with the online retailer PBS Holding, using text embeddings and a nearest neighbor retrieval method. We compare compact encoder models with larger LLM-based models under developmental constraints, including long-tail class distributions and CPU-bound inference. We find that the *multilingual-e5-large* model provides the best Top-1 accuracy and Top-1 balanced accuracy while maintaining a strong CPU efficiency. Some larger LLM-based models achieve competitive results, however, primarily in Top-3 and Top-5 accuracy and for longer product texts. Our evaluation results demonstrate the potential and practical feasibility of applying AI for effective product data enrichment.

Keywords: eClass classification · text embeddings · nearest neighbor retrieval · large language models · e-commerce product data quality

1 Introduction

Product master data management is crucial for product discovery, filtering, and ultimately purchase decisions in e-commerce. In practice, retailers such as our company partner PBS holding operate large and heterogeneous product catalogs with continuously changing products and attributes while heavily relying on vendor-provided information of varying quality. Inaccurate or incomplete textual product descriptions and illustrating images may reduce search quality and can lead to misleading product presentations for customers. These challenges result

© The Author(s), under exclusive license to Springer Nature Switzerland AG 2027
M. Geiß et al. (Eds.): DEXA 2026 Workshops, CCIS 3078, pp. 39–50, 2027.
https://doi.org/10.1007/978-3-032-34202-7_4

in substantial manual effort for data stewards, increased operational costs, and data quality issues. This becomes even more challenging when product information must be maintained consistently across multiple languages, as is the case for our partner. Consequently, PBS Holding is exploring AI based approaches to support scalable product data enrichment and quality assurance under real-world operational constraints.

This paper presents an applied AI use case conducted in cooperation with the online retailer PBS Holding. The scientific contribution focuses on nearest neighbor eClass classification using text embeddings to map new products to the most likely eClass category based on vendor-provided descriptions. We benchmark multiple state-of-the-art models on an internal dataset with ground truth labels and evaluate their classification accuracy and runtime performance.

Section 2 gives an overview of the related work, Sect. 3 presents the methodology and Sect. 4 discusses the results. Finally, Sect. 5 summarizes the findings while Sect. 6 concludes with future directions.

2 Related Work

Automated eClass classification, in literature often referred to as category prediction, is a central e-commerce task [3,8,14,20] and remains challenging under long-tail class distributions and noisy labels, often occurring in real-world product catalogs. Classical supervised approaches can perform well in balanced settings, but class imbalance and label noise are known to reduce robustness and generalization [7,9]. These limitations motivate retrieval-oriented solutions that rely on semantic representations rather than on class-discriminative training. Sentence level embedding models such as *Sentence-BERT* [16] and *LaBSE* [6] established strong semantic matching capabilities, while newer multilingual retrieval models such as the *E5* family [19], *GTE* [22], and *BGE-M3* [13] improved accuracy and cross-lingual robustness. Recent studies further show that LLM-based embedders can provide competitive general purpose representations [18], with strong multilingual variants such as *Qwen3-Embedding*, Llama-Embed-Nemotron, and KaLM-Embedding [1,23,24]. However, results on the Massive Text Embedding Benchmark (MTEB) [12] and its multilingual extension [10] indicate that performance significantly varies across tasks and domains. Consequently, models with high performance on public benchmarks may still underperform on domain-specific taxonomies such as eClass. German product text data is clearly under-represented in pretraining data, which highlights the importance of in-domain evaluation over reliance on public benchmark ratings. As an alternative to embedding-based retrieval, related work also studies direct product understanding using prompting techniques in LLMs. In e-commerce, LLMs have been applied to product categorization in zero-shot settings [17], or through systems that combine fine-tuned and general-purpose models [5]. However, when the number of target categories grows large, generative approaches struggle to produce consistent outputs that are grounded in taxonomy and therefore are subject to hallucinations. Chen et al. [4] address this using a taxonomy-embedded framework that enforces hierarchical consistency across classification levels, but their approach still relies on LLM inference at prediction time. This method faces following constraints: inference costs that scale with catalog size and hallucinations of non-existent categories. These limitations motivate retrieval-based methods that reframe classification as nearest-neighbor search over a fixed, known label space. In our work, predictions are derived exclusively from catalog entries with verified eClass labels, ensuring that outputs stay grounded in the existing taxonomy. This retrieval formulation is related to nearest-neighbor classification over learned representations, which has been shown to handle long-tail distributions effectively by preserving rare-class examples in a non-parametric memory [11]. Unlike parametric classifiers, such approaches do not require retraining when new categories or products are added to the catalog. Beyond text-based approaches, vision-language models (VLMs) that jointly process image and text inputs have also seen rapid development in recent years [21]. More recent work has adapted these capabilities specifically to e-commerce. Nulli et al. [14] present a backbone-agnostic recipe for adapting VLMs to attribute extraction from product images, benchmarking Qwen3-VL, *LLaMA 3.1*, and *Gemma 3* variants on a large set of commercial product listings. Their work focuses on attribute prediction from isolated product images and does not address the eClass classification problem presented in this study.

3 Methodology

3.1 Task Definition

Together with PBS Holding, we aim to automate the assignment of eClass categories from vendor-provided product texts under realistic production constraints. The task is framed as an embedding-based nearest neighbor classification, where new product descriptions are mapped to semantically similar catalog entries with known eClass labels. We evaluate both compact encoder style models and larger LLM-based embedding models to analyze the trade off between classification quality and runtime. A key challenge is the highly imbalanced, long-tail class distribution, with many rare classes and noisy product descriptions.

3.2 Data and Annotation

Applying AI to e-commerce product data faces multiple data challenges. Online product catalogs continuously evolve as new vendors and products are added, while labeled data is limited for fine-grained categories. The overall class distribution is highly imbalanced and long-tailed, containing multiple rare classes, enhancing the challenge for supervised learning. Furthermore, most product descriptions are written in German, whereas many large language models are primarily trained on English text, limiting their effectiveness on data of different language origin.

We use an internal PBS dataset both to generate the embedding catalog and to evaluate classification performance. The ground truth eClass labels were created manually as part of operational processes. The dataset for the embedding catalog generation comprises 272,628 product text entries with corresponding eClass labels spanning 2,492 unique eClasses from 1991 different vendors. 15.13% of entries were excluded due to missing product descriptions. The dataset covers all database entries in the system on 25th of June 2024. Table 1 summarizes the distribution of product entries per class. The class distribution is highly imbalanced as only around 100 classes contain more than 500 entries, while nearly half of the classes have less than 10 entries. The dataset used for classification evaluation includes 20,811 entries covering 871 unique eClasses. 12.71% of entries were excluded due to missing text. This dataset contains all data entered to the system from February 2025 to November 2025. All analyzed product text descriptions are in German. However, eClass classification for additional languages will be integrated into the production system in the future. An example product description and eClass name alongside a product image and an application image are shown in Fig. 1.

TESA Gewebeband Perfect grün
19 mm × 2,75 m (56341-00032-03)
eClass: 27-18-04-01 Klebeband

"Das Vielzweckband mit sehr starker Klebekraft für kleine Reparaturen, zum Fixieren und Bündeln. Extrem belastbar, reißfest, wetter- und wasserfest."

Fig. 1. Example product entry with textual description, eClass, and images.

Table 1. Distribution of product entries per class in the embedding catalog.

Number of entries	1	2–5	6–10	11–20	21–100	101–500	501–1000	1001–14000
Number of classes	433	589	266	255	584	279	64	50

3.3 Models

We evaluate eight multilingual text embedding models covering a wide range of model capacities, from compact encoders to large scale large language based models. The smaller, production oriented models include *multilingual-e5-small* [19], *gte-large* [22],*multilingual-e5-large* [19], *LaBSE* [6], and *bge-m3* [13]. The larger models *Qwen3-Embedding-8B* [23], *llama-embed-nemotron-8b* [1], and *KaLM-Embedding-Gemma3-12B* [24] were specifically included because they rank among the top performers on MTEB multilingual embedding benchmarks [10,12]. Our operational baseline is *multilingual-e5-large*, selected for its favorable trade off between classification quality and inference efficiency under CPU-only hourly batch inference constraints.

3.4 Methodology and Evaluation Metrics

For each model, we first create an embedding catalog by encoding all available product texts from a large product database and store one embedding vector per product entry. During inference, a new product description is embedded with the same model and compared to the catalog vectors using nearest neighbor search in the embedding space with cosine similarity as the distance metric, a common approach in embedding based retrieval [16,19]. All embeddings are L2 normalized before similarity computation. The predicted eClass label is then derived from the labels of the top retrieved neighbors using exact top-k nearest-neighbor search.

Evaluation is conducted on an independent classification dataset. We report Top-1, Top-3, and Top-5 accuracies as the primary evaluation metrics as well as the balanced Top-1, Top-3 and Top-5 accuracies to account for the highly imbalanced and long-tailed class distribution, as conventional Top-k accuracy metrics tend to be dominated by well-represented classes and may overestimate real-world performance. Balanced accuracy on the other hand provides a more robust estimate of model behavior across the full class spectrum as it computes the average recall across all classes, thereby assigning equal weight to both frequent and rare categories [2].

Embedding models derived from large language models are often combined with a separate supervised classifier trained on top for instruction tuning. Because such approaches are commonly considered strong baselines in embedding benchmarks [13], we additionally train a logistic regression classifier on *KaLM-Embedding-Gemma3-12B* embeddings, as this model achieved the highest accuracy among the three larger LLM-based evaluated models. We use *"Represent this text for eClass product classification:"* as our instruction prompt and the scikit-learn implementation of multinomial logistic regression with default L2 regularization [15]. At inference time, predictions are obtained directly from the trained classifier applied to the embedding vectors.

In addition, we analyze the influence of input text length on classification accuracy and report the runtime performance of each model under both CPU and GPU inference settings.

4 Results

4.1 Classification Accuracy

Table 2 summarizes Top-1, Top-3, and Top-5 accuracies related to model size (number of parameters) across all evaluated models, while Table 3 additionally reports the balanced Top-1, Top-3 and Top-5 accuracies to account for the highly imbalanced class distribution. Table 2 shows that *multilingual-e5-large* achieves the strongest overall performance (Top-1: 70.22%, Top-3: 86.88%, Top-5: 90.89%), followed by *multilingual-e5-small* (Top-1: 67.46%), and *KaLM-Embedding-Gemma3-12B* (Top-1: 66.29%).

These results demonstrate the strong practical effectiveness of the E5 family despite its comparatively moderate model size. This behavior is consistent with their training objective based on contrastive learning for retrieval-style semantic matching, which aligns closely with the nearest neighbor classification approach used in this work [19]. In contrast, the larger LLM-based embedding models (8-12B parameter) do not outperform encoder style models in the Top-1 accuracy, despite their strong performance on general-purpose embedding benchmarks. A likely explanation is that these models are optimized for broad semantic understanding and retrieval across diverse domains rather than fine-grained, domain-specific classification task such as our eClass classification use case [1,23,24]. In practice, such models are often paired with a task specific classifier, discussed in Sect. 4.4. Less recent models show noticeably weaker results. In particular, *LaBSE*, which was originally designed for multilingual sentence similarity and translation alignment, achieves only 40.83% Top-1 accuracy. This performance gap likely reflects architectural differences as well as the fact that it relies on comparatively small embedding dimensions [6]. This behavior is also consistent with the text length analysis discussed in Sect. 4.2.

Table 3 further refines the analysis by reporting balanced Top-1, Top-3 and Top-5 accuracies, providing a more robust evaluation under the long-tailed class distribution. As expected, all models exhibit a decline compared to standard Top-k accuracy, confirming that performance is generally higher on frequent classes than on rare ones. For instance, *multilingual-e5-large* Top-1 accuracy decreases from 70.22% to 60.05%, highlighting the difficulty of correctly predicting underrepresented eClass categories. Despite this overall reduction, the ranking of models remains largely consistent, with the *multilingual-e5* family continuing to perform strongly. Notably, *multilingual-e5-large* also achieves the highest balanced Top-1 accuracy (60.05%), indicating that its performance advantage extends beyond frequent classes into the long-tail distribution. However, for balanced accuracies *KaLM-Embedding-Gemma3-12B* achieves the highest balanced Top-3 and Top-5 accuracies (76.20% and 81.64%), suggesting that while some large LLM-based models may not consistently achieve the top prediction, they are still able to retrieve semantically relevant candidates within the top-ranked set. Comparing standard and balanced metrics further reveals model-specific behavior. In particular, *Qwen3-Embedding-8B* and *llama-embed-nemotron-8b* show relatively improved rankings under balanced evaluation, indicating better generalization to rare classes than suggested by Top-1 accuracy alone. In contrast, models such as *gte-large* and *bge-m3* exhibit larger performance drops, suggesting a stronger bias toward frequent classes.

4.2 Classification Accuracy by Text Length

To analyze the influence of product text length on classification accuracy, the evaluation dataset was divided into ten percentile segments based on character count (without white spaces), ranging from short descriptions in segment 1 to long descriptions in segment 10, listed in Table 4. Figure 2 shows the Top-1 accuracy across these segments for selected models. Overall, the majority of models benefit from longer input texts, as additional context can provide stronger semantic signals for embedding similarity. This effect is particularly visible for

Table 2. Top-1, Top-3, and Top-5 accuracies and parameter count.

Model	Parameter	Top-1 Acc. (%)	Top-3 Acc. (%)	Top-5 Acc. (%)
multilingual-e5-small	118M	67.46	85.09	89.17
gte-large	335M	56.01	75.20	80.35
LaBSE	471M	40.83	59.63	66.19
multilingual-e5-large	560M	**70.22**	**86.88**	**90.89**
bge-m3	568M	60.70	78.78	83.71
Qwen3-Embedding-8B	8B	60.66	80.23	85.69
llama-embed-nemotron-8b	8B	61.63	82.11	87.92
KaLM-Embedding-Gemma3-12B	12B	66.29	85.64	89.84

Table 3. Balanced Top-1, Top-3, and Top-5 accuracies and parameter count.

Model	Parameter	Bal. Top-1 Acc. (%)	Bal. Top-3 Acc. (%)	Bal. Top-5 Acc. (%)
multilingual-e5-small	118M	55.94	72.05	77.30
gte-large	335M	47.83	64.15	68.95
LaBSE	471M	35.47	51.98	57.85
multilingual-e5-large	560M	**60.05**	75.15	79.25
bge-m3	568M	50.29	69.27	75.11
Qwen3-Embedding-8B	8B	52.11	72.20	78.63
llama-embed-nemotron-8b	8B	55.64	72.17	77.54
KaLM-Embedding-Gemma3-12B	12B	57.61	**76.20**	**81.64**

Table 4. Range of characters (without white spaces) for each segment.

Percentile Segment	1	2	3	4	5	6	7	8	9	10
Min char	4	52	69	90	121	170	237	315	418	603
Max char	51	68	89	120	169	236	314	417	602	2901

gte-large, whose Top-1 accuracy increases from roughly 47% for short inputs to over 60% for longer descriptions. Differences between encoder-based and LLM-based embedding models are also apparent. The best model *multilingual-e5-large* maintains a consistently strong performance across all text lengths, whereas LLM-derived embeddings improve more noticeably with longer inputs. In particular, the *KaLM-Embedding-Gemma3-12B* model gradually increases in performance and slightly surpasses the *multilingual-e5-large* for the longest two text segments. This behavior likely reflects the larger context windows and train-

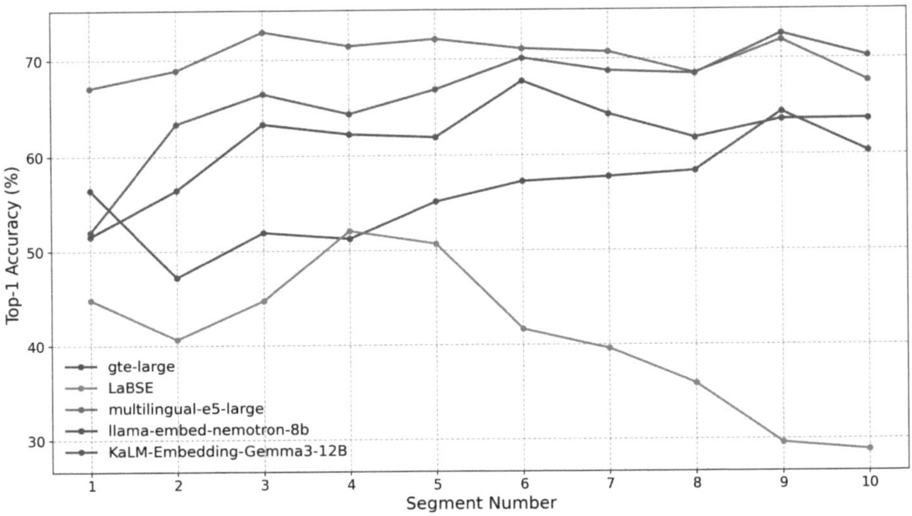

Fig. 2. Top-1 accuracy by text length across 10 segments.

ing regimes of modern LLM-based embedding models, optimized for processing longer sequences [18]. In contrast, *LaBSE* shows decreasing performance as input length increases, with a Top-1 accuracy dropping from about 45% to below 30%. This decline likely results from its comparatively smaller embedding capacity and older training objectives [6].

4.3 Runtime Performance

In addition to classification quality, runtime performance is a key criterion for production deployment. Table 5 reports the average runtime per single eClass classification call (embedding product text, nearest neighbor search, and retrieval of top-5 classes), computed over 100 individual calls and averaged separately for CPU and GPU execution. Experiments were conducted on a compute cluster equipped with an AMD EPYC 7643 processor (48 cores, 2.3 GHz) and NVIDIA A100 GPUs. As expected, runtime strongly scales with model size. CPU latency differs by almost two orders of magnitude between compact encoders and large LLM based embedders (e.g., 0.86 s for *multilingual-e5-large* vs. 21.38 s for *KaLM-Embedding-Gemma3-12B*). Although GPU acceleration mitigates most of this overhead (e.g., 0.16 s for *multilingual-e5-large* vs. 0.20 s for *KaLM-Embedding-Gemma3-12B*), their inference latency remains higher than that of mid scale encoder models. Consequently, the trade off between model size and inference cost is a central factor for model selection in production systems. This is particularly relevant in the operational PBS environment, where CPU only constraints enforce hourly batch inference.

Table 5. Average runtime per single eClass classification call.

Model	Parameter	CPU (s)	GPU (s)
multilingual-e5-small	118M	0.26	0.10
gte-large	335M	0.92	0.18
LaBSE	471M	0.39	0.11
multilingual-e5-large	560M	0.86	0.16
bge-m3	568M	0.84	0.16
Qwen3-Embedding-8B	8B	15.11	0.20
llama-embed-nemotron-8b	8B	13.81	0.20
KaLM-Embedding-Gemma3-12B	12B	21.38	0.20

4.4 Training a Classifier

Table 6 shows the accuracy and balanced accuracy results for the *KaLM-Embedding-Gemma3-12B* model with and without a trained classifier on top. Across all evaluated metrics the added classifier leads to a small but consistent decrease in performance. In terms of Top-1, Top-3, and Top-5 accuracy, the classifier-based variant underperforms the direct embedding similarity approach by approximately 0.8–1.3 percentage points, with a comparable drop in the balanced accuracy metrics as well. These findings indicate that the embedding space is already highly structured for nearest-neighbor-based inference, and that imposing a global linear decision boundary may slightly distort the intrinsic metric geometry rather than improve class separability. Further investigations using alternative classification heads as well as experiments across the other embedding models would be useful to assess whether this behavior generalizes across architectures and training regimes in the future (see Sect. 5).

Table 6. Accuracy and balanced accuracy for *KaLM-Embedding-Gemma3-12B* with and without an additional classifier.

Model	Top-1 Acc. (%)	Top-3 Acc. (%)	Top-5 Acc. (%)
KaLM-Embedding-Gemma3-12B	66.29	85.64	89.84
KaLM-Embedding-Gemma3-12B Classifier	65.47	84.37	88.63
Model	Bal. Top-1 Acc. (%)	Bal. Top-3 Acc. (%)	Bal. Top-5 Acc. (%)
KaLM-Embedding-Gemma3-12B	57.61	76.20	81.64
KaLM-Embedding-Gemma3-12B Classifier	55.95	74.51	79.85

5 Conclusion

The E5 family still remains a strong and practical baseline for embedding-based eClass classification, even with the emergence of more recent and significantly larger LLM-based embedding models. Across both standard and balanced evaluation metrics, the *multilingual-e5-large model* achieves the best overall Top-1 accuracies, all at substantially lower computational cost compared to multi-billion parameter models. At the same time, the *KaLM-Embedding-Gemma3-12B* model shows particularly strong performance in the balanced setting, achieving the highest Top-3 and Top-5 balanced accuracies, indicating advantages in handling class imbalance and capturing minority classes. Nevertheless, this improvement does not translate into Top-1 performance, which remains the most critical metric for a classification scenario.

Preliminary experiments with instruction-based prompting and an additional classifier heads did not lead to performance improvements for the *KaLM-Embedding-Gemma3-12B* model. This suggests that classification-oriented transformations may distort the embedding space in a way that is detrimental to similarity-based retrieval tasks. Further investigations using alternative classification heads, as well as experiments across additional embedding models, would be useful to assess whether these findings generalize across architectures and training regimes.

Overall, considering both accuracy and efficiency, the results indicate that larger LLM-based embedding models, with or without additional classifier heads, do not provide a clear practical advantage over well-established smaller models such as the E5 family in this setting. The architecture and contrastive training objectives of the E5 family, specifically designed for semantic retrieval, make these models particularly well suited to nearest-neighbor classification, also for shorter product text length. Furthermore, they provide significantly better CPU efficiency, which is critical for real-world production environments where latency and resource constraints are key considerations.

6 Outlook

Future work will extend the eClass classification workflow towards multi-modal embedding based retrieval that incorporates product images in addition to text. Such approaches could be particularly valuable when textual descriptions are missing, incomplete or incorrectly specified in vendor provided descriptions. A central challenge in this context is the alignment of visual and textual representations within a shared retrieval space while maintaining stable behavior and predictable performance across heterogeneous product categories.

In addition to the current work, we aim to explore automated attribute extraction from product texts in order to populate structured fields for downstream applications such as faceted search and improved catalog navigation. Leveraging recent advances in generative AI, we further plan to introduce automatic product description generation conditioned on product images and the predicted eClass label. This functionality could also support multilingual description generation, reducing manual translation effort while maintaining consistent terminology.

Finally, we envision an interactive shopping assistant that exploits the enriched product database, enabling customers to articulate product requirements in natural language and efficiently retrieve suitable items, while providing clear references to the underlying catalog entries.

Acknowledgments. The research reported in this paper has been partly funded by BMIMI, BMWET, and the State of Upper Austria in the frame of the SCCH competence center INTEGRATE [(FFG grant no. 892418)] part of the FFG COMET Competence Centers for Excellent Technologies Programme.

References

1. Babakhin, Y., et al.: Llama-embed-nemotron-8b: a universal text embedding model for multilingual and cross-lingual tasks. arXiv preprint arXiv:2511.07025 (2025)
2. Brodersen, K.H., Ong, C.S., Stephan, K.E., Buhmann, J.M.: The balanced accuracy and its posterior distribution. In: 2010 20th International Conference on Pattern Recognition, pp. 3121–3124. IEEE (2010)
3. Cevahir, A., Murakami, K.: Large-scale multi-class and hierarchical product categorization for an e-commerce giant. In: Proceedings of COLING 2016, the 26th International Conference on Computational Linguistics: Technical Papers, pp. 525–535 (2016)
4. Chen, S., et al.: Leveraging taxonomy and LLMs for improved multimodal hierarchical classification. In: Proceedings of the 31st International Conference on Computational Linguistics (COLING 2025), pp. 6244–6254. Association for Computational Linguistics, Abu Dhabi, UAE (2025). https://doi.org/10.18653/v1/2025.coling-main.417
5. Cheng, Z., et al.: E-commerce product categorization with LLM-based dual-expert classification paradigm. In: Proceedings of the 1st Workshop on Customizable NLP: Progress and Challenges in Customizing NLP for a Domain, Application, Group, or Individual (CustomNLP4U), pp. 294–304. Association for Computational Linguistics, Miami, Florida, USA (2024). https://doi.org/10.18653/v1/2024.customnlp4u-1.22
6. Feng, F., Yang, Y., Cer, D., Arivazhagan, N., Wang, W.: Language-agnostic BERT sentence embedding. In: Proceedings of the 60th Annual Meeting of the Association for Computational Linguistics (Volume 1: Long Papers), pp. 878–891 (2022)
7. Frénay, B., Verleysen, M.: Classification in the presence of label noise: a survey. IEEE Trans. Neural Netw. Learn. Syst. **25**(5), 845–869 (2014)
8. Gao, D.: Deep hierarchical classification for category prediction in E-commerce system. In: Proceedings of the 3rd Workshop on e-Commerce and NLP, pp. 64–68 (2020). https://doi.org/10.18653/v1/2020.ecnlp-1.10
9. Gao, X., et al.: A comprehensive survey on imbalanced data learning. Front. Comput. Sci. **20**(11) (2026). https://doi.org/10.1007/s11704-025-50274-7
10. Hugging Face: Mteb leaderboard. Hugging Face (2026). https://huggingface.co/spaces/mteb/leaderboard

11. Long, A., et al.: Retrieval augmented classification for long-tail visual recognition. In: Proceedings of the IEEE/CVF Conference on Computer Vision and Pattern Recognition (CVPR), pp. 2235–2245 (2022)
12. Muennighoff, N., Tazi, N., Magne, L., Reimers, N.: MTEB: massive text embedding benchmark. In: Proceedings of the 17th Conference of the European Chapter of the Association for Computational Linguistics, pp. 2014–2037 (2023)
13. Multi-Granularity, M.L.M.F.: M3-embedding: multi-linguality, multi-functionality, multi-granularity text embeddings through self-knowledge distillation. arXiv preprint arXiv:2402.03216 (2024)
14. Nulli, M., et al.: Adapting vision-language models for E-commerce understanding at scale. arXiv preprint arXiv:2602.11733 (2026). https://arxiv.org/abs/2602.11733
15. Pedregosa, F., et al.: Scikit-learn: machine learning in Python. J. Mach. Learn. Res. **12**, 2825–2830 (2011)
16. Reimers, N., Gurevych, I.: Sentence-BERT: sentence embeddings using Siamese BERT-networks. In: Proceedings of the 2019 Conference on Empirical Methods in Natural Language Processing and the 9th International Joint Conference on Natural Language Processing (EMNLP-IJCNLP), pp. 3982–3992 (2019)
17. Roumeliotis, K.I., Tselikas, N.D., Nasiopoulos, D.K.: LLMs for product classification in e-commerce: a zero-shot comparative study of GPT and Claude models. Nat. Lang. Process. J. **11**, 100142 (2025). https://doi.org/10.1016/j.nlp.2025.100142
18. Tao, C., et al.: LLMs are also effective embedding models: an in-depth overview (2025)
19. Wang, L., Yang, N., Huang, X., Yang, L., Majumder, R., Wei, F.: Multilingual E5 text embeddings: a technical report. arXiv preprint arXiv:2402.05672 (2024)
20. Yu, W., Sun, Z., Liu, H., Li, Z., Zheng, Z.: Multi-level deep learning based e-commerce product categorization. In: eCOM@ SIGIR (2018)
21. Zhang, J., Huang, J., Jin, S., Lu, S.: Vision-language models for vision tasks: a survey. IEEE Trans. Pattern Anal. Mach. Intell. (TPAMI) (2024)
22. Zhang, X., et al.: MGTE: generalized long-context text representation and reranking models for multilingual text retrieval. In: Proceedings of the 2024 Conference on Empirical Methods in Natural Language Processing: Industry Track, pp. 1393–1412 (2024)
23. Zhang, Y., et al.: Qwen3 embedding: advancing text embedding and reranking through foundation models. arXiv preprint arXiv:2506.05176 (2025)
24. Zhao, X., et al.: KaLM-embedding-v2: superior training techniques and data inspire a versatile embedding model. arXiv preprint arXiv:2506.20923 (2025)

Scalable Federated Adaptation of Foundation Models Using Flower on Kubernetes-Based HPC Infrastructure: A Technical Report

Batuhan Bencik, Mathias Brucker, Mohit Kumar, Roman Rainer, and Alexander Valentinitsch(✉)

Software Competence Center Hagenberg GmbH, Softwarepark 32a, 4232 Hagenberg, Austria

alexander.valentinitsch@scch.at

Abstract. We present a scalable federated learning system for adapting vision-language foundation models using Low-Rank Adaptation (LoRA) on a Kubernetes-based HPC infrastructure. Clients perform local fine-tuning on private data, while a central server aggregates LoRA updates to maintain a global model. We evaluate the system on a real-world multi-site dataset and compare federated and centralized training. Results show that LoRA enables efficient adaptation with reduced communication overhead, while federated learning improves generalization across heterogeneous data. Federated learning achieves comparable localization performance while significantly reducing wall-clock training time through parallel execution on multiple GPUs. These results demonstrate a practical and scalable approach for privacy-preserving adaptation of foundation models in distributed and high-performance computing environments.

Keywords: Federated Learning · Foundation Models · Low-Rank Adaptation (LoRA) · Kubernetes · High-Performance Computing (HPC) · Flower Framework

1 Introduction

Foundation models have fundamentally transformed modern artificial intelligence by enabling strong generalization across a wide range of tasks through large-scale pretraining on diverse data [17]. Vision-language models such as Florence-2 unify several capabilities within one architecture, including image

The research reported in this paper has been supported by the AI Mission Austria Flagship Project FAIR-AI [(FFG grant no. 904624)] and BMIMI, BMWET, and the State of Upper Austria in the frame of the SCCH competence center INTEGRATE [(FFG grant no. 892418)] part of the FFG COMET Competence Centers for Excellent Technologies Programme.

© The Author(s), under exclusive license to Springer Nature Switzerland AG 2027
M. Geiß et al. (Eds.): DEXA 2026 Workshops, CCIS 3078, pp. 51–62, 2027.
https://doi.org/10.1007/978-3-032-34202-7_5

captioning, object detection, and visual grounding [21]. This reduces the need for specialized models and large labeled datasets. Despite these advances, adapting such models to domain-specific or privacy-sensitive settings remains a significant challenge [6].

Traditional centralized training approaches, which aggregate all data in a single location, often yield strong performance but are frequently impractical in real-world scenarios due to privacy constraints, regulatory requirements, and the cost of transferring large-scale datasets [6]. These limitations are particularly pronounced in distributed and high-performance computing (HPC) environments. Federated learning (FL) offers a compelling alternative by enabling multiple clients to collaboratively train a shared global model without exchanging raw data; instead, clients compute local updates and communicate model parameters to a central server. Frameworks such as Flower simplify the deployment of such systems [2]. Nevertheless, federated learning introduces several challenges, including increased communication overhead, system heterogeneity, and slower convergence, which are particularly pronounced for large foundation models [24].

To address these limitations, parameter-efficient fine-tuning techniques have emerged as a promising approach. In particular, Low-Rank Adaptation (LoRA) reduces the number of trainable parameters by decomposing weight updates into low-rank matrices while keeping the pretrained backbone frozen [7]. This significantly lowers memory usage and communication cost, making LoRA particularly well-suited for federated environments where bandwidth and computational resources are constrained.

In this work, we present a scalable federated learning system for adapting foundation models using LoRA on a Kubernetes-based HPC infrastructure. Federated clients are deployed as containerized pods that perform local training, while a central server aggregates the LoRA updates and maintains the global model. This setup allows efficient use of resources, scales across heterogeneous GPU nodes, and supports reproducible distributed training.

We compare our approach with centralized training, focusing on convergence, training time, communication efficiency, and model performance. We also explore how the choice of LoRA rank affects the trade-off between accuracy and communication cost. Overall, the results suggest that the system remains competitive while keeping data local and lowering communication overhead, which is particularly useful in distributed and privacy-sensitive environments.

Contributions

- A scalable federated learning framework for foundation model adaptation using Flower on a Kubernetes-based HPC infrastructure.
- Integration of LoRA for parameter-efficient fine-tuning, enabling reduced communication and memory requirements.
- A fully containerized client-server pipeline supporting reproducible and flexible distributed training.
- An empirical comparison of federated and centralized training, including analysis of scalability and LoRA rank effects.

2 Related Work

Federated learning (FL) has emerged as a key paradigm for decentralized model training, enabling multiple clients to collaboratively learn a shared model without exchanging raw data. Early work by [13] introduced the Federated Averaging (FedAvg) algorithm, which remains a widely used baseline for aggregating client updates. Subsequent research has highlighted challenges such as communication efficiency, client heterogeneity, and convergence instability in non-i.i.d. settings [8]. These issues become more pronounced when scaling FL to large foundation models [17], motivating the need for efficient training and communication strategies.

Recent advances in federated learning frameworks, such as Flower [2], provide flexible and extensible infrastructures for deploying FL systems across heterogeneous environments. Flower enables seamless integration with deep learning libraries (e.g., PyTorch, and Hugging Face Transformers) and supports both simulation and real-world distributed deployments, making it suitable for large-scale experimentation. Compared to earlier frameworks, it emphasizes modularity and scalability, which are critical for adapting foundation models in practical settings [5].

To address the computational and communication challenges of training large models, parameter-efficient fine-tuning methods have gained significant attention. Low-Rank Adaptation (LoRA) [7] introduces a decomposition of weight updates into low-rank matrices, allowing only a small subset of parameters to be trained while freezing the base model. This approach has been successfully applied to large language and vision models, significantly reducing memory usage and communication costs [7]. In federated settings, LoRA is particularly advantageous, as it minimizes the amount of data exchanged between clients and the server while maintaining competitive performance [10,19,23], with recent advances achieving up to 79% communication reduction [11].

From a systems perspective, container orchestration platforms such as Kubernetes have become the de facto standard for managing distributed machine learning workloads in cloud and high-performance computing (HPC) environments. Kubernetes enables scalable deployment, resource isolation, and fault tolerance through containerized services, making it well-suited for federated learning scenarios involving multiple clients and dynamic resource allocation [16]. Prior work has demonstrated the effectiveness of Kubernetes for distributed training pipelines [20]; however, its integration with federated learning frameworks and parameter-efficient fine-tuning methods remains relatively underexplored [14].

Finally, foundation models have opened up new possibilities, but they also make distributed training more difficult because of their size, generality, and the way data can vary across clients [4,24]. Although centralized fine-tuning is still widely used, more recent work has begun to explore federated adaptation as a way to protect privacy and limit the movement of sensitive data [17]. Even so, much of the current literature still treats system scalability and model efficiency as separate problems [15,22]. Our work brings these aspects together by combining federated learning, LoRA-based adaptation, and Kubernetes orchestration in a single scalable framework.

3 Methodology

Problem Formulation. We consider a federated learning (FL) setup with N distributed clients. Each client $i \in \{1, \ldots, N\}$ holds a private local dataset D_i with size $|D_i|$. The overall dataset is defined as $D = \bigcup_{i=1}^{N} D_i$, with total size $|D| = \sum_{i=1}^{N} |D_i|$. The goal is to train a shared global model W without requiring clients to exchange their raw data [8,13].

Training proceeds in communication rounds. At each round t, the server sends the current global model $W^{(t)}$ to all clients. Each client performs local optimization using its own data and returns an update to the server. The global objective can be written as a weighted sum of local losses:

$$\min_{W} \sum_{i=1}^{N} \frac{|D_i|}{|D|} \mathcal{L}_i(W),$$

where $\mathcal{L}_i(W)$ denotes the loss on client i.

To make training large foundation models more efficient in this setting, we adopt Low-Rank Adaptation (LoRA) [7]. Instead of updating all model parameters, the base model W remains fixed, and only low-rank update matrices are trained. For each client i, the adapted model is expressed as:

$$W_i' = W + \Delta W_i, \quad \Delta W_i = A_i B_i,$$

where $A_i \in \mathbb{R}^{d \times r}$ and $B_i \in \mathbb{R}^{r \times d}$, with $r \ll d$.

Under this formulation, each client optimizes only its local LoRA parameters:

$$\min_{A_i, B_i} \mathcal{L}_i(W + A_i B_i),$$

while the base model W remains unchanged. This significantly reduces the number of trainable parameters and the amount of data exchanged during training.

After local training, the server aggregates the client updates using weighted averaging:

$$\Delta W^{(t)} = \sum_{i=1}^{N} \frac{|D_i|}{|D|} \Delta W_i^{(t)},$$

and updates the global model accordingly:

$$W^{(t+1)} = W^{(t)} + \Delta W^{(t)}.$$

By restricting updates to low-rank matrices, the communication cost is reduced from $\mathcal{O}(d^2)$ to $\mathcal{O}(d \cdot r)$, making the approach practical for large-scale foundation models in federated environments.

System Architecture. The proposed system is implemented as a containerized federated learning pipeline based on the Flower framework and deployed on a Kubernetes-orchestrated HPC cluster [15]. The overall architecture follows a client-server design, as illustrated in Fig. 1. The server and all clients run as isolated Kubernetes pods, improving portability, reproducibility, and operational control.

Federated Clients: Each client is deployed as an independent pod with dedicated compute resources (CPU/GPU). Clients perform local fine-tuning of the foundation model using private datasets. Only LoRA parameters are updated during training, while the base model remains frozen. After each communication round, clients transmit LoRA updates to the server.

Federated Server: The server is responsible for coordinating the federated learning process. It initializes the global model, distributes it to clients, aggregates received LoRA updates, and computes the updated global model. Additionally, it evaluates model performance on a shared test dataset and logs system metrics such as accuracy, training time, and communication cost.

Communication Workflow: Training is carried out in synchronous communication rounds using the Flower framework [2]. In each round, the server distributes the current global model to all clients. The clients perform local training on their private data and compute LoRA updates, which are then sent back to the server. The server aggregates these updates to produce the next global model. By exchanging only LoRA parameters, the communication overhead per round is significantly reduced. This workflow corresponds to the bidirectional communication shown in Fig. 1.

Kubernetes Orchestration: Kubernetes manages deployment, scaling, and resource allocation across the HPC cluster. Each client and server runs in a containerized environment, enabling reproducibility, fault tolerance, and dynamic scaling. This setup allows efficient utilization of heterogeneous compute resources.

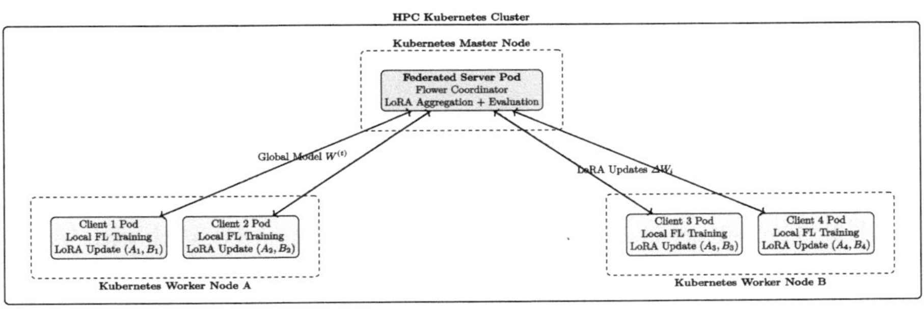

Fig. 1. HPC-based federated learning architecture using Flower and LoRA on a Kubernetes cluster. Clients are distributed across worker nodes, while a central server coordinates aggregation and evaluation.

4 Experiments

Deployment Setup. The federated learning system is deployed on a Kubernetes-based high-performance computing (HPC) cluster with four worker nodes, each hosting one federated client. This one-to-one mapping ensures resource isolation and prevents interference between concurrent training workloads.

Clients are deployed as containerized Kubernetes pods with dedicated compute resources, including CPU, memory, and a single GPU. The federated server is deployed as a lightweight service without fixed resource constraints, as it primarily handles orchestration and aggregation.

Communication between server and clients is implemented using the Flower framework over gRPC, leveraging Kubernetes internal networking. Training proceeds in synchronous federated learning rounds: the server distributes the global model, clients perform local updates using Low-Rank Adaptation (LoRA), and only the resulting adapter parameters are transmitted back. This significantly reduces communication overhead compared to full model synchronization.

All components are containerized and deployed via Kubernetes, ensuring reproducibility and portability. The detailed deployment configuration is summarized in Table 1. This setup reflects a typical distributed FL configuration on HPC infrastructure, balancing resource isolation and efficient utilization.

Table 1. Deployment configuration of the federated learning system on a Kubernetes-based HPC cluster. Clients are deployed as GPU-enabled pods, while the server components handle orchestration and aggregation.

Component	Configuration	Description
Client Pods (x4)	1 per Kubernetes node	One-to-one mapping between clients and nodes ensuring isolation
CPU (Client)	4–8 cores (request/limit)	Dedicated CPU allocation per client pod
GPU (Client)	1 × NVIDIA GPU (A100/V100)	Allocated via NVIDIA device plugin for each client
Memory (Client)	16–64 GB RAM	Node-dependent memory allocation
Server Pod (Flower SuperLink)	No explicit CPU/GPU limits	Lightweight orchestration and coordination layer
Server Application Job	No explicit CPU/GPU limits	Handles aggregation and evaluation logic
Communication	gRPC via Flower framework	Internal Kubernetes ClusterIP networking
Deployment Platform	Kubernetes HPC Cluster	Container orchestration for distributed training

Dataset and Preprocessing. We use a real-world multi-site object detection dataset derived from the waste-sorting use case introduced in our previous work [19]. The dataset is distributed across four clients (A-D), each with its own local training and test split, reflecting a realistic federated learning setting.

The data is stored as JSONL files with annotations formatted for Florence-2 object detection, including object labels and bounding box coordinates. In total, the dataset contains 18 object classes and 1000 annotated instances. The class distribution is imbalanced, with some classes occurring significantly more frequently than others, introducing non-IID conditions across clients.

For preprocessing, images and annotations are converted into the Hugging Face `datasets` format and tokenized using the Florence-2 processor. To increase sample diversity, we apply a cropping strategy around ground-truth bounding boxes, which generates multiple training samples from a single image.

After preprocessing, the dataset consists of 5311 training samples and 1166 test samples across all clients (see Table 2). For centralized experiments, we additionally construct a pooled dataset by combining all client training splits, which serves as an upper-bound reference for performance comparison.

Table 2. Processed dataset statistics after preprocessing and cropping.

Client	Train Samples	Test Samples
Client A	1330	281
Client B	1327	280
Client C	1424	313
Client D	1230	292
Total	**5311**	**1166**

Experimental Setup. We evaluate the effectiveness of parameter-efficient federated fine-tuning of a vision-language foundation model, focusing on (i) the impact of LoRA rank, (ii) the comparison between federated and centralized training, and (iii) system-level efficiency.

All experiments are conducted using the Florence-2 model adapted with Low-Rank Adaptation (LoRA). The dataset is distributed across clients to simulate a realistic federated setting, while centralized training on the pooled dataset serves as a performance upper bound.

Federated training follows a synchronous protocol, where all clients participate in each round and perform one local epoch of training. The server aggregates LoRA updates to obtain the global model. For comparability, centralized and federated training are evaluated using the same budget of 25 epochs/rounds.

Table 3. Experimental configuration.

Parameter	Value
Framework	Flower
Number of Clients	2, 3, 4
LoRA Rank	4, 8, 16
Training Rounds	25
Environment	Kubernetes HPC Cluster

Table 3 summarizes the key experimental parameters. We evaluate LoRA ranks $r \in \{4, 8, 16\}$ and vary the number of participating clients. All experiments are executed on a Kubernetes-based HPC cluster with dedicated GPU resources per client.

5 Experimental Results

LoRA Rank Analysis. Increasing the LoRA rank improves performance, with test loss decreasing from 3.70 ($r = 4$) to 3.51 ($r = 8$) and 3.43 ($r = 16$). The improvement from $r = 8$ to $r = 16$ is relatively small, indicating diminishing returns, while training time increases only marginally. Rank analysis completed on the small subset of the processed dataset. (see Table 4).

Table 4. Performance and training time comparison of different LoRA ranks.

LoRA Rank	Eval Loss	Test Loss	Time (min)
4	3.7687	3.6993	23.38
8	3.5575	3.5119	23.69
16	**3.4650**	**3.4338**	24.96

Local vs Federated Training. Federated learning improves generalization compared to local training. Independently trained models show variability, with test losses between 2.10 and 2.18, whereas the federated model achieves a more stable performance of 1.81 by leveraging data across clients (see Table 5).

Table 5. Local vs federated learning comparison.

Method	Setting	Test Loss	Total Time (min)
Local Training	Client A	2.10	404.57
Local Training	Client B	2.18	411.71
Local Training	Client C	2.12	469.27
Local Training	Client D	2.16	390.11
Federated Learning	Global Model	1.81	494.30

Impact of Number of Clients. Increasing the number of clients reduces test loss, while localization performance remains relatively stable. However, communication cost increases notably, indicating a trade-off between model quality and system efficiency (see Table 6).

Table 6. Impact of the number of clients on model performance and system cost.

# Clients	Test Loss	mIoU	mAP@50	Train	Comm.	Total Time (min)
2 (A,B)	2.01	0.697	0.011	304.3	144.3	448.6
3 (A,B,C)	1.85	0.697	0.010	317.3	170.9	488.2
4 (A,B,C,D)	1.81	0.676	0.014	317.8	176.5	494.3

Federated vs Centralized Training. Under the same training budget, federated learning achieves comparable localization performance while reducing wall-clock time through parallel client training on multiple GPUs, at the cost of communication overhead (Table 7).

Table 7. Comparison of centralized and federated training.

Method	Rounds/Epochs	Test Loss	mIoU	mAP@0.5	Time (min)
Centralized	25 epochs	1.73	0.685	0.0181	2371
Federated	25 rounds	1.81	0.676	0.0141	494

6 Discussion

The experimental results provide several key insights into parameter-efficient federated fine-tuning of foundation models.

The impact of LoRA rank confirms the effectiveness of parameter-efficient adaptation for large models. As shown in Table 4, increasing the rank improves performance, while gains diminish beyond moderate values such as $r = 8$. This suggests that compact adapter representations capture most task-relevant information, consistent with prior work on LoRA and PEFT methods [7] [22] [12]. Evaluating this behavior on larger and more diverse datasets remains an important direction for future work.

The comparison between local and federated training highlights the benefit of collaborative learning. As shown in Table 5, federated learning produces a more robust global model compared to independently trained local models, which exhibit high variability due to heterogeneous data distributions [13] [8]. However, the dataset itself introduces additional challenges, as it contains complex scenes with multiple objects and background noise, which can negatively impact model generalization, especially for locally trained models. Extending this setup to asynchronous federated learning could further improve efficiency in such heterogeneous environments.

From a system perspective, federated learning introduces non-trivial engineering challenges. Orchestrating multiple client processes as distributed Kubernetes pods requires careful coordination, particularly when managing synchronization across training rounds. In addition, GPU scheduling for multiple concurrent federated clients is challenging in shared HPC environments, where resource contention and allocation policies can significantly affect runtime and stability. These practical constraints contribute to communication and orchestration overhead, although parallel execution across multiple GPUs substantially reduces wall-clock training time.

The scalability analysis further reinforces this trade-off. As shown in Table 6, increasing the number of clients improves performance due to greater data diversity [3], while communication cost becomes the dominant factor in total runtime. This highlights the importance of efficient communication strategies, as well as robust orchestration mechanisms to handle increasing system complexity. Improving fault tolerance and scheduling in Kubernetes-based deployments is therefore an important next step.

The convergence analysis shows that federated learning achieves comparable localization performance while substantially reducing wall-clock training time through parallel execution across multiple GPUs. (Table 7). Nevertheless, decentralized updates, non-iid data distributions, and partial information exchange remain fundamental challenges for federated optimization [8,13].

From a systems perspective, these results highlight the importance of communication efficient techniques. LoRA reduces transmitted parameters while maintaining competitive performance [1,10,11]. Integrating and benchmarking with frameworks such as NVFlare would further strengthen the evaluation and provide a standardized comparison baseline.

Overall, the results highlight key trade-offs between performance, communication cost, system complexity, and convergence speed. While LoRA-based federated fine-tuning provides a practical and scalable solution, real-world deployment requires careful consideration of infrastructure constraints, data quality, and system robustness.

This work opens several promising directions for future research. First, we plan to explore asynchronous federated learning and adaptive client selection strategies to better handle heterogeneous resources and mitigate straggler effects, while maintaining scalability and robustness. Second, we will investigate additional communication-efficient approaches beyond LoRA. In particular, kernel-based and space-folding representations could be used to compress high-dimensional model updates into compact, task-relevant summaries [9]. Combining these techniques with LoRA adapters may further reduce communication overhead while preserving model performance. Finally, we plan to study alternative communication backends, such as NVIDIA NVFlare [18], as a transport layer within the existing Flower-based architecture. This would allow a more systematic analysis of communication efficiency, scalability, and system-level trade-offs across different federated learning runtimes.

7 Conclusion

In this work, we presented a federated learning system for adapting a vision-language foundation model using LoRA on a Kubernetes-based HPC setup. We evaluated the impact of LoRA rank, compared federated and local training, and analyzed the effect of varying the number of participating clients.

The results show that LoRA enables efficient fine-tuning with reduced communication overhead. Federated learning improves generalization compared to local training, but introduces additional communication and coordination costs. Increasing the number of clients further enhances performance, while also increasing communication overhead. Federated learning achieves performance comparable to centralized training while reducing overall wall-clock training time through distributed execution.

Overall, combining LoRA with federated learning provides a practical approach for adapting foundation models in distributed and privacy-sensitive environments, balancing performance and efficiency.

References

1. Babakniya, S., et al.: Slora: federated parameter efficient fine-tuning of language models, arXiv preprint (2023). arXiv:2308.06522
2. Beutel, D.J., et al.: Flower: A friendly federated learning research framework. arXiv preprint arXiv:2007.14390 (2020)
3. Bommasani, R., et al.: On the opportunities and risks of foundation models. arXiv preprint arXiv:2108.07258 (2021)
4. Fan, T., et al.: Ten challenging problems in federated foundation models. IEEE Trans. Knowl. Data Eng. (2025)
5. Flower Labs: Flower: A Friendly Federated AI Framework. https://github.com/flwrlabs/flower (Apr 2026), version 1.13.0
6. Hatfaludi, C.A., Serban, A.: Foundational models and federated learning: survey, taxonomy, challenges and practical insights. PeerJ Comput. Sci. **11**, e2993 (2025)
7. Hu, E.J., et al.: Lora: low-rank adaptation of large language models. ICLR **1**(2), 3 (2022)
8. Kairouz, P., McMahan, H.B.: Advances and open problems in federated learning. Foundat. Trends Mach. Learn. **14**(1–2), 1–210 (2021)
9. Kumar, M., et al.: Operator-theoretic framework for gradient-free federated learning, arXiv preprint arXiv:2512.01025 (2025)
10. Kuo, K., Raje, A., Rajesh, K., Smith, V.: Federated lora with sparse communication. arXiv preprint arXiv:2406.05233 (2024)
11. Liu, H., et al.: Ecolora: communication-efficient federated fine-tuning of large language models. In: Proceedings of the 2025 Conference on Empirical Methods in Natural Language Processing, pp. 20743–20757 (2025)
12. Mangrulkar, S., Gugger, S., Debut, L., Belkada, Y., Paul, S., Bossan, B.: Peft: State-of-the-art parameter-efficient fine-tuning methods (2022). https://github.com/huggingface/peft
13. McMahan, B., Moore, E., Ramage, D., Hampson, S., y Arcas, B.A.: Communication-efficient learning of deep networks from decentralized data. In: Artificial intelligence and statistics. pp. 1273–1282. PMLR (2017)
14. Parra-Ullauri, J.M., et al.: Privacy preservation in kubernetes-based federated learning: A networking approach. In: IEEE INFOCOM 2023-IEEE Conference on Computer Communications Workshops (INFOCOM WKSHPS), pp. 1–7. IEEE (2023)

15. Parra-Ullauri, J.M., et al.: kubeflower: a privacy-preserving framework for kubernetes-based federated learning in cloud-edge environments. Futur. Gener. Comput. Syst. **157**, 558–572 (2024)
16. Pham, K.Q., Kim, T.: Elastic federated learning with kubernetes vertical pod autoscaler for edge computing. Futur. Gener. Comput. Syst. **158**, 501–515 (2024)
17. Ren, C., et al.: Advances and open challenges in federated foundation models. IEEE Commun. Surv. & Tutorials (2025)
18. Roth, H.R., et al.: Supercharging federated learning with flower and nvidia flare. In: International Workshop on Trustworthy Federated Learning, pp. 36–45. Springer (2024). https://doi.org/10.1007/978-3-031-82240-7_3
19. Valentinitsch, A., et al.: Collaborative trustworthy foundation model framework: an environmental sustainability use-case to detect contamination objects in organic waste streams. In: International Conference on Database and Expert Systems Applications, pp. 30–44. Springer (2025). https://doi.org/10.1007/978-3-032-02003-1_3
20. Woźniak, M., Sochat, V.: Kubecon: Enabling hpc and ml workloads with the latest kubernetes job features. KubeCon EU (Apr 2023). https://software.llnl.gov/news/2023/04/21/kubecon/, [web:132]
21. Xiao, B., et al.: Florence-2: advancing a unified representation for a variety of vision tasks. In: Proceedings of the IEEE/CVF Conference on Computer Vision and Pattern Recognition, pp. 4818–4829 (2024)
22. Yang, Y., Long, G., Lu, Q., Zhu, L., Jiang, J., Zhang, C.: Federated low-rank adaptation for foundation models: A survey, arXiv preprint arXiv:2505.13502 (2025)
23. Yi, L., Yu, H., Wang, G., Liu, X., Li, X.: pfedlora: model-heterogeneous personalized federated learning with lora tuning, arXiv preprint arXiv:2310.13283 (2023)
24. Zhuang, W., Chen, C., Lyu, L.: When foundation model meets federated learning: motivations, challenges, and future directions, arXiv preprint arXiv:2306.15546 (2023)

Synthetic Data for Training and Evaluation in Industrial Settings

Johannes Zellinger[1(✉)], Hrishikesh Gupta[1], Andreas Winkler[2], Christian Kohlberger[2], Clara Schachner[1], and Florian Kromp[1]

[1] Software Competence Center Hagenberg GmbH, Hagenberg, Austria
johannes.zellinger@scch.at
[2] Trumpf SE + Co. KG, Stuttgart, Germany

Abstract. High quality data gathering and labeling for model training and testing is a challenging and time consuming task. In industrial settings, generating diverse datasets–including sample variations, edge cases, and environmental factors–is challenging. These efforts are often constrained by production limitations, high costs, and error rates. To tackle these problems, during the course of the CVRL COMET project, a pipeline to produce synthetic image datasets with controllable variations, automatic labeling and additional meta-information was created. The pipeline was initiated and implemented in cooperation with the vision team at Trumpf SE + Co. KG. The project's focus was to create automatic tests and benchmarks for existing pattern matching used in production setups as well as provision of new datasets to test new methods and algorithms in terms of feasibility and weaknesses. In addition, the domain gap between real-world and synthetic data, and its effects, was investigated and mitigation techniques were explored. The close collaboration with project partners enabled a valuable exchange of expertise, benefiting both academic research and industrial applications. The proposed system enables partners to virtually test new production setups, reducing risks and costs before implementing real-world adaptations.

Keywords: Synthetic Data · Data Generation · Data Representation · Blender

1 Introduction

The use of machine learning methods covers a wide range of applications in industrial settings, including robotics, vision-based quality control or process optimization among others. Applied models must be trained and tested on high quality data in order to produce reliable outputs. However, in industrial production, providing the necessary amounts of sampled data covering all expected variations of the entire population is expensive and might be restricted by additional factors, such as data privacy concerns of customers. Additionally, rapidly changing products or production lines can necessitate retraining with new datasets

© The Author(s), under exclusive license to Springer Nature Switzerland AG 2027
M. Geiß et al. (Eds.): DEXA 2026 Workshops, CCIS 3078, pp. 63–74, 2027.
https://doi.org/10.1007/978-3-032-34202-7_6

frequently. A solution to solve this problem is to apply foundational zero- or few-shot models, utilizing knowledge from web-scale training on downstream tasks without the need to train for a particular application with large annotated datasets. The development of specialized industrial applications however can be challenging when the appearance of data samples is highly specific, not covered by large-scale datasets crawled from the internet, a strategy commonly used to train foundational models. However, in production scenarios, geometric data of a product is often available in the form of CAD drawings or 3D models and could be used to guide training.

This paper presents an applied use case for automatically generate synthetic data generation from available CAD data in order to be used for training of AI models as well as to provide benchmarks and rigorous testing for existing algorithms. We further explore the current state of the art for data generation methods and review, whether these can be applied to the task at hand.

2 Related Work

Procedural 3D Assets and Materials. Recent work in procedural 3D content generation has shown increasing interest in using node-based representations and foundational models to automate the creation and editing of virtual assets. Blender has been particularly attractive in this context, due to its Python API and node system programmatic control over geometry, lighting, and physically based materials. Huan et al. [15] demonstrates this by using vision-language models to iteratively edit Blender programs according to text or image-based user intent, including procedural materials, geometry, and lighting. Similarly, Gu et al. [11] introduces a benchmark for evaluating foundational models on Blender editing tasks such as object placement, lighting, procedural material editing, blend-shape manipulation, and geometry editing. However, these studies also indicate that even recent vision-language systems still struggle with reliable and precise 3D graphics editing, in particular when low-level visual properties must be controlled through modifications in program or node. A parallel line of work focuses specifically on procedural material generation. Guerrero et al. [12] models procedural materials as editable, resolution-independent node graphs and further generates graph nodes, parameters, and edges using a multi-stage transformer model. Subsequently, the work done by Li et al. [18] improves image-conditioned procedural material generation by using reinforcement learning to optimize material parameters with respect to appearance similarity rather than only parameter-space supervision. *VLMaterial* [19] further adapts large vision-language models to generate Blender procedural material programs from input images, producing editable shader node graphs.

Geometry generation methods can be categorized into methods, that generate new data from text prompts utilizing SDS-based optimization [3,30,39] and reconstruction methods, creating meshes from input frames covering multiple viewpoints [13,22] or point clouds [13]. Models following the latter gained increased popularity after the introduction of the usage of *neural radiance fields (NeRf)* [26] which was further optimized by MÃijller et al. [27]. This trend was continued by using *Gaussian splatting* [17], both providing a way to capture and reconstruct 3D data from input view-pose pairs.

Texture and materials reconstruction methods have been a prominent research topic since as a result of *NeRFs*, which enables storage of light transport information across scenes. *NeILF* [35] follows the general idea of *NeRF* and represents scene illumination as a neural incident light field, encoding spatially-varying incoming radiance at any scene point and direction via an MLP. Building on this foundation, *NeILF++* combines light estimation with geometry reconstruction by introducing inter-reflection constraints linking the incident and outgoing light fields, thus enforcing physical consistency. Finally, *PBR-NeRF* adds additional physics-based losses, which allows improved disentanglement of physically based materials from light transport across the scene. Such reconstruction based methods work on a per-instance basis, trying to capture objects observed in the input data. However, generating new instances of materials based on the same general properties is not possible with such approaches. This is addressed by the work of Henlzler et al. [14] which presents a method to train a model on flash images of materials with non-repeating features. The model learns the natural distribution of the material and can produce new outputs from queries with random noise as part of the input. Recently, image generation methods, especially based on diffusion models have been used to generate textures and materials for 3D objects [7,16].

3 Objectives and Problem Statement

3.1 Overview

The collaborative project between Trumpf SE + Co. KG. and the CVRL team at SCCH within the COMET program aims to achieve knowledge transfer within the scientific and industrial partners as well as strengthening Austria's position as a key location for business and research. The goal is to develop innovative solutions for key challenges such as industrial automation.

Trumpf's Austrian branch focuses on industrial robots for automatic bending of metal sheets. The company's product is a bending cell consisting of a secured area, where industrial robots process stacks of metal sheets with task specific bending tools. Grabbing the next sheet from a stack is an automated process, driven and controlled by a vision system detecting the top sheet and transferring its pose information to the robot. This tasks presents several hard to solve problems, such as items on a stack having identical geometry and similar surfaces, metal surfaces being reflective and changing environments in different production sites among others.

The task solved within the cooperation is to improve precision as well as time consumption of the detection and pose estimation of metal sheets given its geometry as CAD information and images from the camera system within the cell.

3.2 Production Setup

Production setup consists of different arrangements of one or multiple industrial cameras producing grayscale images. Cameras are calibrated and the images are rectified using OpenCV [2] functions. With a multitude of different production setups, cameras can be positioned stationary or fixed to the robots head. Depending on the application, the system can take a single image or stack multiple images with different exposure values to produce pictures with a high dynamic range. A shape detector aims to find translation and rotation of a metallic product part within the given image by analyzing corresponding CAD data of its contour.

In order to find the best possible set of parameters and to be able to dynamically adapt to changing environments, it is essential to test this system for a wide range of conditions. Creating a dataset manually, covering all possible circumstances appearing within a production setup is a hard and expensive task and thus, options for synthetic data generations are commonly explored. Synthetic data generation is a fast evolving topic, especially with the emergence of diffusion-based image generation models [32] and models aligning text to images [31]. Unfortunately, even with the introduction of ControlNet [37] to guide stable diffusion models, allowing to constrain image generation to predefined image aspects, it is still not possible to determine all resulting properties of the data synthesized by such models. However, precise control over dimensions, materials and object shapes of the produced data is essential to solve the task at hand.

A review of state of the art 3D geometry generation methods show similar shortcomings. Additionally, while for foundational models for image synthesis can be trained on web-scale datasets, such vast training sets are not available in 3D space. Most models explore strategies to project knowledge from models trained on 2D image data into 3D space [20,23,30,36,39]. While some large scale datasets exist [5,6,38] these are based on web-scraped 3D models from popular hosting sites for 3D modelers containing mostly artistic data showing inconsistent materials, scales and shapes that are not based on real world objects. There are curated subsets such as Objaverse++ [21] as well as datasets of 3D scanned real-world objects [9,28], but these are much smaller in size and only applicable for use cases that consist the limited range of object categories present within them.

A possible strategy to retain necessary control over generated data while also generating random variation in materials surfaces, poses and environmental conditions is to employ a programmable rendering pipeline making use of procedural textures for physical-based rendering (PBR) materials. In Sects. 4.2 and 4.3 we share a detailed description of implementation details of the developed pipeline.

4 Methodology

4.1 Contour Detection with Real World Data

To generate meaningful benchmarking datasets, a comparison with real world data has been conducted. A real world dataset was provided by Trumpf consisting of 1120 samples covering 14 different shapes and a variety of material properties. Additionally, the dataset is split into different stack types, describing how the pose of individual metal sheets varies within one stack. Each sample is composed of 4 raw images, lit from different light sources and combined into fused images. Multiple ways of processing the raw images into single images are used and the different results are provided in the datasets. We test the Halcon software on this dataset to have a baseline for benchmarks on synthetic data. The algorithm returns position in x and y direction, rotation angle as well as scale. The values are based on the loaded geometry data, using the origin to provide the position and rotation. The scale is given as $mm/pixel$. We calculate the mean absolute error for each of the returned metrics over all instances per shape. Figure 3 visualizes the performance of the matcher using parameters from empirical tests.

4.2 Render Pipeline

Blender 3D [1] is commonly used as a tool for data generation, processing and visualization in scientific contexts [8,10,25]. It's python API allows for seamless integration into a data generation pipeline. Node-based procedural geometry and texture editors enable the creation of an unlimited amount of data variation while keeping the overall desired structure and distribution properties from a small set of settable parameters.

Within the rendering pipeline, procedural materials are not only used for metal sheet rendering, but in addition for background objects such as concrete or wood materials for the floor and euro-pallets, respectively. Overall lighting is simulated via imported high dynamic range images from PolyHaven [29]. Different lighting setups are achieved by adding light sources into the scene and changing properties such as positions, intensity, color and spread among others via script. 3D Geometry of the metal sheets is generated by importing 2D CAD data, adding faces to the elements with *Constrained Delaunay Triangulation* [4] and extruding to the desired thickness. Additionally, small bevels and geometric irregularities are added to the mesh to simulate realistic production errors.

Finally, the rendering script logs required label data, such as object poses, camera pose and parameters or lighting conditions. Additionally we can optionally output additional metadata and training labels such as object masks, depth images, scene point clouds or anomaly heat maps offering a wide range of options for future applications.

4.3 Procedural PBR Materials

The rendering pipeline makes use of procedurally generated textures for all instances that need controlled variations. While high quality *PBR* datasets like *MatSynth* [33] do exist, it is of high importance to have an unbounded number of variations that can be observed on the surfaces of the datasets geometries. This is in particular true if the resulting dataset might be used to train detection models in the future. Otherwise, models would potentially simply learn surface structures instead of the targeted knowledge.

The procedural textures used in the pipeline are based on a combination of *Worley* [34] and *Fractal Brownian Motion* [24] noise provided as parameterize-able nodes in Blender. By empirically combining multiple such texture with different parameters and scales and using them as input for color, metallicness, roughness and normals of the shader distributions resembling any non-repeating materials can be generated. Figure 1 illustrates the result of such materials as well as intermediate outputs to control surface values.

(a) (b) (c)

Fig. 1. Examples showing use of procedural textures. Note that the resulting textures of image (a) and image (b) differ completely per pixel, while keeping the same overall structure. Image (c) displays the texture controlling deeper bumps in the concrete material of the floor.

5 Project Results

In this section, we showcase qualitative results. We demonstrate that the render pipeline allows to have fine-grained control over various visual aspects that can be expected in real world applications, including artifacts due to environmental circumstances, bright light sources, lens or camera sensor degradation among others.

5.1 Qualitative Results

Figure 2 illustrates generated synthetic images as well as introduced anomalies that can incorporated. With the proposed setup, we can simulate a multitude of problematic artifacts, which can be expected during production setups, especially for systems running over a long time period. Examples of simulated artifacts include, but are not limited to

- Over- or under exposure: extreme lighting conditions such as direct sun or spotlight illumination as well as wrong camera settings like high or low exposure times or high sensitivity settings.
- Camera sensor degradation: over the course of time, camera sensors can degrade resulting in image patches having varying intensity values and additional sensor noise.
- Camera lens artifacts: artifacts include lens distortions, chromatic aberration, lens dispersion, depth of field, camera sensor offsets or lens contamination and scratching among others.
- Geometric errors of the matched objects: achieved by adding procedural noise to vertex positions as well as by animating the objects and storing armature information. An example of a geometric error can be observed in Figure 2.

5.2 Quantitative Results

We evaluate and compare results of the applied matcher between rendered and real datasets. Figure 3 demonstrates that the resulting metrics are comparable, although the matcher overall performs slightly better on the rendered dataset. This is due to the nature of the real dataset which combines multiple images, thereby creating wider contour lines for the top sheets while the rendering setup in this case was generated in a single pass.

In addition, we show that the matching performance can be influenced by introducing artifacts. By controlling the strength of the artifacts, threshold values can be found, where the matcher has a strong falloff in precision. Figure 4 illustrates this by adding lens dispersion to the renders and testing the matcher on the resulting images.

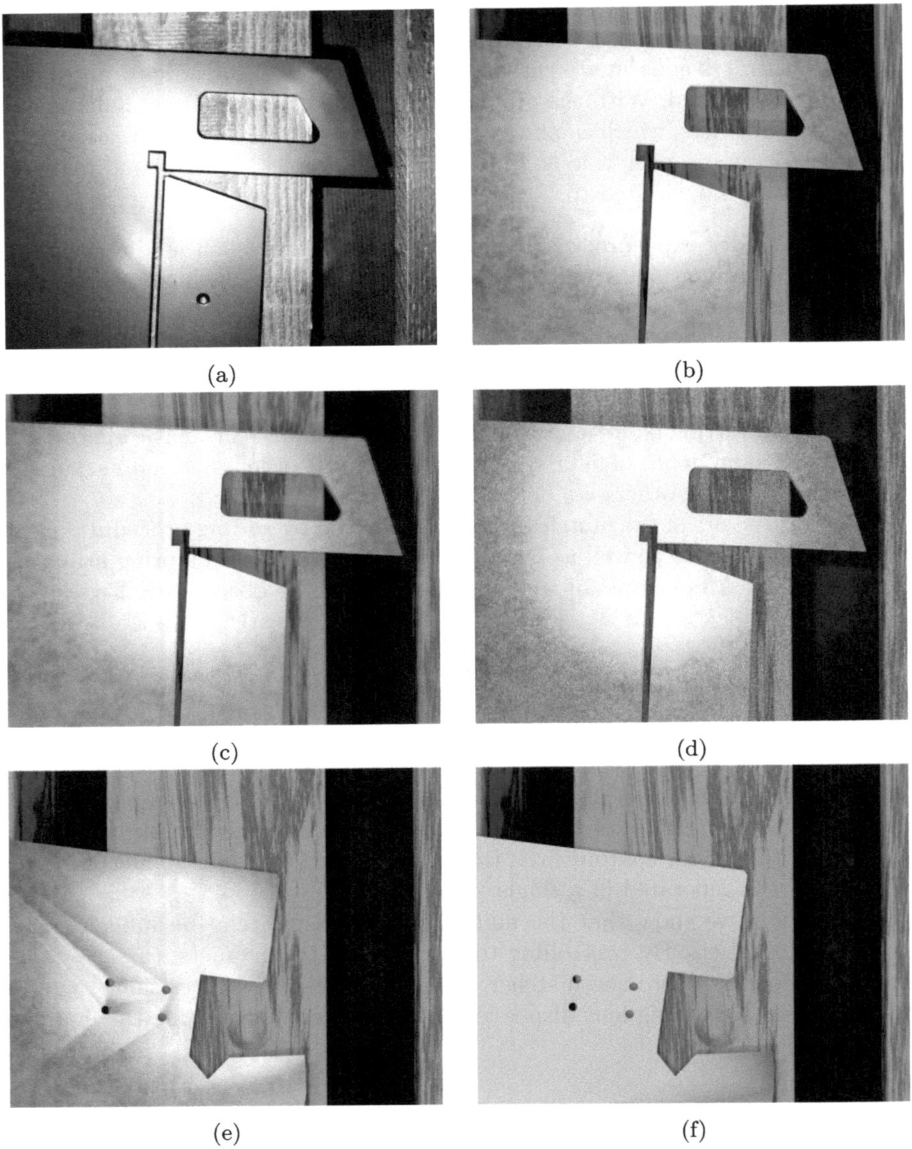

Fig. 2. Example showing a comparison between a real world sample (a) and a rendered example (b). The middle row shows example of simulated anomalies like chromatic aberration (c) and high sensor noise (d). The bottom displays capabilities to introduce geometric artifacts by bending the objects in certain areas (e) along with a corresponding heatmap (f) of the bent areas which can be used as label.

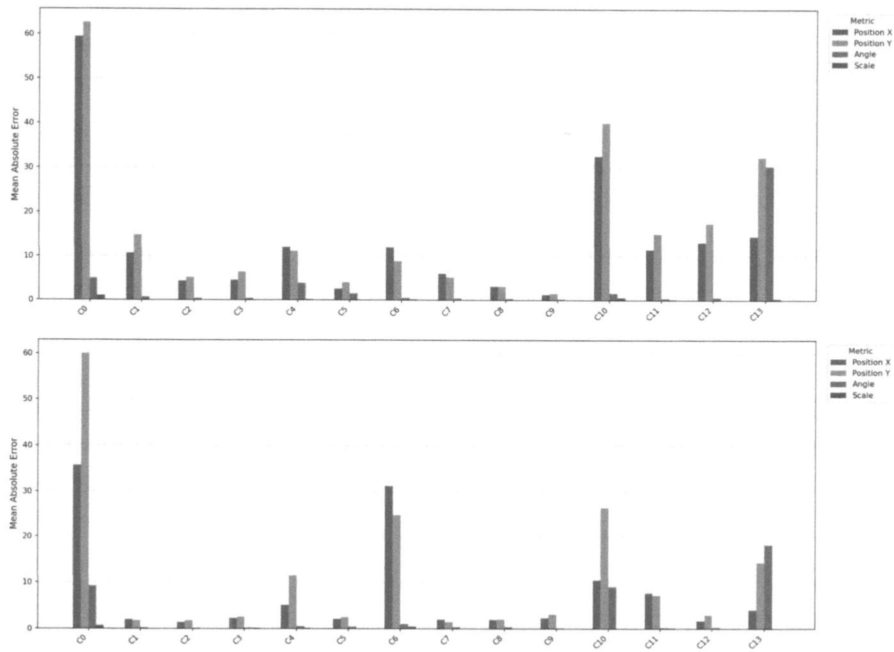

Fig. 3. Mean absolute error of the Halcon matcher applied to real world (top) and rendered (bottom) data across different metal sheet shapes and metrics.

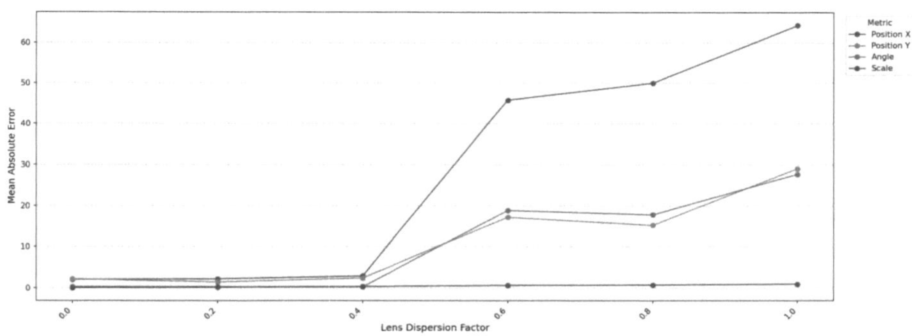

Fig. 4. Graph showing decreased performance of the matcher in relation to increasing lens dispersion factor.

6 Conclusion and Outlook

The rendering pipeline presented in this document provides a wide range of options to benchmark existing matching algorithms and the possibility to provide training data for future machine learning models. The availability of perfect ground truth labels as well as additional meta-information without the need for manual labor enables the generation of high-quality datasets with fine-grained control over produced samples. Future work and research will focus on two main areas:

Automation of Procedural Material Generation. The lack of control over generated materials offer ample research options to combine automated image generation with the advantages of procedural methods. Creating procedural materials with realistic properties is a challenging and time-consuming task.

Domain Gap Minimization. While the proposed pipeline offers a well defined way to generate test and benchmarking data for rule-based algorithms, preliminary analysis revealed differences between real-world and rendered images in latent space of a range of popular feature-extractors like *CLIP* [31]. This is most likely due to domain gaps in the generated synthetic data as compared to real world data which could lead to decreased performance of models trained on this data. Further analysis and research on domain gap mitigation could offer increased performance for future downstream applications.

Acknowledgments. The research reported in this paper has been partly funded by BMIMI, BMWET, and the State of Upper Austria in the frame of the SCCH competence center INTEGRATE [(FFG grant no. 892418)] part of the FFG COMET Competence Centers for Excellent Technologies Programme.

References

1. Blender Online Community: Blender - a 3D modelling and rendering package. Blender Foundation, Stichting Blender Foundation, Amsterdam (2026). http://www.blender.org
2. Bradski, G.: The OpenCV Library. Dr. Dobb's Journal of Software Tools (2000)
3. Chen, Z., Wang, F., Wang, Y., Liu, H.: Text-to-3d using gaussian splatting. In: Proceedings of the IEEE/CVF Conference on Computer Vision and Pattern Recognition, pp. 21401–21412 (2024)
4. Chew, L.P.: Constrained delaunay triangulations. In: Proceedings of the Third Annual Symposium on Computational Geometry, pp. 215–222 (1987)
5. Deitke, M., et al.: Objaverse-xl: a universe of 10m+ 3d objects. Adv. Neural. Inf. Process. Syst. **36**, 35799–35813 (2023)
6. Deitke, M., et al.: Objaverse: a universe of annotated 3d objects. In: Proceedings of the IEEE/CVF Conference on Computer Vision and Pattern Recognition, pp. 13142–13153 (2023)
7. Deng, K., et al.: Flashtex: fast relightable mesh texturing with lightcontrolnet. In: European Conference on Computer Vision, pp. 90–107. Springer (2024)
8. Denninger, M., et al.: Blenderproc. arXiv preprint arXiv:1911.01911 (2019)
9. Downs, L., et al.: Google scanned objects: a high-quality dataset of 3d scanned household items. In: 2022 International Conference on Robotics and Automation (ICRA), pp. 2553–2560. IEEE (2022)
10. Gros, A., et al.: Microscopy nodes: versatile 3d microscopy visualization with blender. EMBO Rep. 1–17 (2026)

11. Gu, Y., Huang, I., Je, J., Yang, G., Guibas, L.: Blendergym: Benchmarking foundational model systems for graphics editing (2025)
12. Guerrero, P., Hašan, M., Sunkavalli, K., Měch, R., Boubekeur, T., Mitra, N.J.: Matformer: a generative model for procedural materials. ACM Trans. Graph. **41**(4), 1–12 (2022). https://doi.org/10.1145/3528223.3530173
13. Held, J., et al.: Meshsplatting: Differentiable rendering with opaque meshes. arXiv (2025)
14. Henzler, P., Deschaintre, V., Mitra, N.J., Ritschel, T.: Generative modelling of brdf textures from flash images, arXiv preprint arXiv:2102.11861 (2021)
15. Huang, I., Yang, G., Guibas, L.: Blenderalchemy: editing 3d graphics with vision-language models (2024)
16. Huang, X., Wang, T., Liu, Z., Wang, Q.: Material anything: generating materials for any 3d object via diffusion. In: Proceedings of the Computer Vision and Pattern Recognition Conference, pp. 26556–26565 (2025)
17. Kerbl, B., Kopanas, G., Leimkühler, T., Drettakis, G., et al.: 3d gaussian splatting for real-time radiance field rendering. ACM Trans. Graph. **42**(4), 139–1 (2023)
18. Li, B., et al.: Procedural material generation with reinforcement learning. ACM Trans. Graph. **43**(6), 1–14 (2024). https://doi.org/10.1145/3687979
19. Li, B., et al.: Vlmaterial: procedural material generation with large vision-language models. In: International Conference on Learning Representations (2025)
20. Lin, C.H., et al.: Magic3d: high-resolution text-to-3d content creation. In: Proceedings of the IEEE/CVF Conference on Computer Vision and Pattern Recognition, pp. 300–309 (2023)
21. Lin, C., et al.: Objaverse++: curated 3d object dataset with quality annotations. In: Proceedings of the IEEE/CVF International Conference on Computer Vision, pp. 6813–6822 (2025)
22. Liu, M., et al.: Meshformer: high-quality mesh generation with 3d-guided reconstruction model. arXiv preprint arXiv:2408.10198 (2024)
23. Long, X., et al.: Wonder3d: single image to 3d using cross-domain diffusion. In: Proceedings of the IEEE/CVF Conference on Computer Vision and Pattern Recognition, pp. 9970–9980 (2024)
24. Mandelbrot, B.B., Van Ness, J.W.: Fractional brownian motions, fractional noises and applications. SIAM Rev. **10**(4), 422–437 (1968)
25. Mehl, L., Schmalfuss, J., Jahedi, A., Nalivayko, Y., Bruhn, A.: Spring: a high-resolution high-detail dataset and benchmark for scene flow, optical flow and stereo. In: Proceedings of the IEEE/CVF Conference on Computer Vision and Pattern Recognition, pp. 4981–4991 (2023)
26. Mildenhall, B., Srinivasan, P.P., Tancik, M., Barron, J.T., Ramamoorthi, R., Ng, R.: Nerf: representing scenes as neural radiance fields for view synthesis. Commun. ACM **65**(1), 99–106 (2021)
27. Müller, T., Evans, A., Schied, C., Keller, A.: Instant neural graphics primitives with a multiresolution hash encoding. ACM Trans. Graph. (TOG) **41**(4), 1–15 (2022)
28. Pan, X., et al.: Aria digital twin: a new benchmark dataset for egocentric 3d machine perception. In: Proceedings of the IEEE/CVF International Conference on Computer Vision, pp. 20133–20143 (2023)
29. Poly Haven: Poly haven – the public 3d asset library (2026). https://polyhaven.com Accessed 20 April 2026
30. Poole, B., Jain, A., Barron, J.T., Mildenhall, B.: Dreamfusion: text-to-3d using 2d diffusion, arXiv preprint arXiv:2209.14988 (2022)

31. Radford, A., et al.: Learning transferable visual models from natural language supervision. In: International Conference on Machine Learning, pp. 8748–8763. PmLR (2021)
32. Song, J., Meng, C., Ermon, S.: Denoising diffusion implicit models, arXiv preprint arXiv:2010.02502 (2020)
33. Vecchio, G., Deschaintre, V.: Matsynth: a modern pbr materials dataset. In: Proceedings of the IEEE/CVF Conference on Computer Vision and Pattern Recognition, pp. 22109–22118 (2024)
34. Worley, S.: A cellular texture basis function. In: Proceedings of the 23rd Annual Conference on Computer Graphics and Interactive Techniques, pp. 291–294 (1996)
35. Yao, Y., et al.: Neilf: neural incident light field for physically-based material estimation. In: European Conference on Computer Vision, pp. 700–716. Springer (2022). https://doi.org/10.1007/978-3-031-19821-2_40
36. Yi, T., et al.: Gaussiandreamer: fast generation from text to 3d gaussians by bridging 2d and 3d diffusion models. In: Proceedings of the IEEE/CVF Conference on Computer Vision and Pattern Recognition, pp. 6796–6807 (2024)
37. Zhang, L., Rao, A., Agrawala, M.: Adding conditional control to text-to-image diffusion models. In: Proceedings of the IEEE/CVF International Conference on Computer Vision, pp. 3836–3847 (2023)
38. Zhang, Y., Zhang, L., Ma, R., Cao, N.: Texverse: a universe of 3d objects with high-resolution textures. arXiv preprint arXiv:2508.10868 (2025)
39. Zhao, Z., et al.: Hunyuan3d 2.0: scaling diffusion models for high resolution textured 3d assets generation. arXiv preprint arXiv:2501.12202 (2025)

Optimisation of Industrial Production with AI Algorithms

Simulation of a Drone-Suspended Pendulum System

Felix Wachter(✉) and G. Ulrich

University of Applied Sciences Kempten, Kempten, Germany
{felix.wachter,ulrich.goehner}@hs-kempten.de

Abstract. A coupled nonlinear simulation framework for a drone carrying a payload suspended via rope is presented. The tethered configuration introduces pendulum-like dynamics coupled to the drone's motion. The framework is intended to support the development of strategies for safe trajectory-following with collision avoidance.

The payload is modeled as a rigid pendulum with a holonomic length constraint enforced via Lagrange multipliers and stabilized using the Baumgarte method, ensuring numerical consistency while preserving physically correct rope forces. The quadrotor model is calibrated up to the attitude-control level, enabling direct specification of thrust magnitude and direction consistent with state-of-the-art autopilot interfaces, while retaining actuator limitations and operating close to the physical origin of the pendulum dynamics.

Simulations demonstrate that the model reproduces physically plausible system behavior, validating its suitability as a basis for control design. System identification experiments confirm frequency-dependent coupling between drone and payload, including tension loss phenomena and nonlinear interaction effects.

The presented framework establishes a foundation for future research into trajectory tracking in the presence of pendulum dynamics, fast and safe payload placement during swing motion, and integrated collision avoidance under simultaneous consideration of pendulum motion, drone trajectory, and actuator constraints.

Keywords: UAV · Sling compensation · Simulation · System Description · Drone Dynamics · Parameter Identification

1 Introduction

Scalable applications of drone technology are emerging as a rapidly growing domain, particularly within the transportation sector. In the context of transportation, a key challenge arises regarding how goods are transferred to their final drop-off location once the designated planar coordinates have been reached. Current commercially available drones transport payloads rigidly attached to the airframe [2], while alternative approaches include winch-based or tethered delivery systems [9] and payload drop mechanisms, with or without parachutes [7]. Each approach is associated with distinct trade-offs.

© The Author(s), under exclusive license to Springer Nature Switzerland AG 2027
M. Geiß et al. (Eds.): DEXA 2026 Workshops, CCIS 3078, pp. 77–89, 2027.
https://doi.org/10.1007/978-3-032-34202-7_7

Landing the Drone: **Pros:** Full control of an approximately rigid system until touchdown; no additional hardware required. **Cons:** Time-intensive; increased collision risk and potential hazard to people due to exposed rotors; high energy demand; potential loss of communication in urban canyons, vegetated areas, or complex terrain.

Payload Drop (Free Fall/Parachute): **Pros:** Fast and energy-efficient; no control required during release. **Cons:** No post-release control; requires predefined drop zones; risk of entanglement with vegetation or infrastructure; no active collision avoidance during descent.

Winch (Tethered Payload): **Pros:** Combines aspects of both approaches; no landing site required (only a deployment area); reduced risk to people due to absence of low-altitude rotors; enables hovering above the operation zone. **Cons:** Partially controllable pendulum dynamics; increased operation time during lowering; challenging release timing under residual oscillations; requires damping prior to payload release.

The present work is motivated by a use case in which a robotic system for tree planting is attached via a tether of at least 30 m in length to a heavy-lift drone equipped with a winch. [10] In the DraAuf project, a DJI FlyCart30 drone is considered. The introduction of the tether leads to coupled nonlinear dynamics, in particular pendulum-like oscillations and potential loss of tether tension, which significantly influence system behavior during payload deployment. Due to simplified flight authorization constraints, the use case is represented in this work by a small-scale model implemented using a DJI Matrice 210 platform, which, due to its common manufacturer, provides comparable interfaces, logging capabilities, and control possibilities.

Fattori et al. [3] address the challenges associated with ground-tethered UAV systems and discuss overlapping physical aspects relevant to this work, such as tether force dynamics and control design under the influence of additional tether-induced forces. Mahony et al. [5] demonstrate how to estimate drone behavior and describe various effects influencing the drone's kinematics. Nishiyama et al. [6] investigate a crane operation support system with damping control, based on model predictive control for a double-pendulum model. However, their approach does not consider regime-dependent effects such as ballistic free-fall dynamics if the cable tension collapses, nor the limitations imposed by the finite thrust capability of the aerial platform, which constrain the achievable counteracting forces.

Sreenath et al. [8] present a geometric control design for a quadrotor UAV with a cable-suspended load, formulated coordinate-free with ideal actuation assumed at the moment level. However, real autopilot interfaces expose only high-level commands such as thrust magnitude and direction, while actuator limits and motor dynamics introduce frequency-dependent effects that influence the coupled pendulum behavior and cannot be neglected in a physically accurate simulation. Furthermore, Cartesian coordinates offer a more transparent physical interpretation of the rope force as a measurable constraint force, facilitating model validation against experimental data, at the cost of requiring explicit enforcement of the length constraint.

The central research question addressed in this work is: **How can a tethered drone–payload system be modeled with sufficient physical fidelity while enabling experimental calibration to serve as a reliable basis for the evaluation of control strategies for collision-safe payload delivery?**

To answer this question, a suitable simulation model must be developed that captures the key dynamics of the coupled system, including the interactions between the drone, the tether, and the payload. Accurate modeling is a prerequisite for required development of control strategies, evaluating stability characteristics, and optimizing payload transport under varying operating conditions, and collision-aware trajectory control.

2 Methodology

We present a closed-loop simulation framework for a quadrotor carrying a payload on a rope in flight, with exemplary length of $l = 30\,\text{m}$.

The full system state is $\mathbf{s} = [\mathbf{s}_d^\top, \mathbf{s}_p^\top]^\top \in \mathbb{R}^{23}$, comprising a 17-state drone model and a 6-state pendulum model. Both subsystems are coupled through the rope force.

2.1 Pendulum Model

The payload position and velocity in world coordinates form the pendulum state $\mathbf{s}_p = [\mathbf{p}, \dot{\mathbf{p}}]^\top \in \mathbb{R}^6$. Letting $\mathbf{r} = \mathbf{p}_p - \mathbf{p}_d$ denote the rope vector from the drone attachment point to the payload, the equation of motion is seen in Eq. 1

$$m_p\ddot{\mathbf{p}} = m_p\mathbf{g} - c_d\|\dot{\mathbf{p}}\|\dot{\mathbf{p}} + \mathbf{F}_{rope}, \tag{1}$$

where $\mathbf{g} = [0, 0, -g]^\top$ and $c_d = \frac{1}{2}\rho C_w A$ denotes the lumped aerodynamic drag coefficient. The rope force $\mathbf{F}_{rope} = -T\hat{\mathbf{r}}$, with $\hat{\mathbf{r}} = \mathbf{r}/\|\mathbf{r}\|$, is calculated from a rigid rope model via a holonomic constraint $\|\mathbf{r}\| = l$ via a Lagrange multiplier T see Eq. 2

$$T = m_{\text{eff}}\left(\frac{\|\dot{\mathbf{r}}_\perp\|^2}{\|\mathbf{r}\|} + \hat{\mathbf{r}}\cdot(\mathbf{a}_p - \mathbf{a}_d) + 2\alpha\,(\dot{\mathbf{r}}\cdot\hat{\mathbf{r}}) + \beta^2(\|\mathbf{r}\| - l)\right), \tag{2}$$

where m_{eff} reduced mass, $\dot{\mathbf{r}}_\perp$ tangential relative velocity, $\mathbf{a}_p$ and $\mathbf{a}_d$ free payload and drone accelerations, and (α, β) Baumgarte stabilization gains [1]. If $T \leq 0$ the rope is slack and the payload is in ballistic free fall.

The rope force returned to external callers is the physical tension $\mathbf{F}_{rope,\text{phys}} = -T_{\text{phys}}\,\hat{\mathbf{r}}$, where T_{phys} excludes the Baumgarte correction terms. This ensures that numerical constraint stabilization affects only the internal pendulum dynamics and does not introduce fictitious forces into the coupled system equations or the slack condition $T_{\text{phys}} \leq 0$. Consequently, the rigid model avoids the numerical stiffness introduced by large spring constants k in a spring damper model.

2.2 Quadrotor Dynamics

The drone state is $\mathbf{s}_d = [\mathbf{p}_d, \mathbf{v}_d, \mathbf{q}, \boldsymbol{\omega}, T_1, T_2, T_3, T_4]^\top \in \mathbb{R}^{17}$, where $\mathbf{q} \in \mathbb{R}^4$ is the unit quaternion in scalar-last convention $[x, y, z, w]^\top$ and $\boldsymbol{\omega}$ is the body-frame angular velocity. It's Translation is described in Eq. 3

$$m_d \ddot{\mathbf{p}}_d = \mathbf{R}(\mathbf{q})\,\mathbf{e}_z \textstyle\sum_i T_i + m_d \mathbf{g} - c_{w,d}\|\mathbf{v}_d\|\mathbf{v}_d - \mathbf{F}_{rope}, \tag{3}$$

where $\mathbf{R}(\mathbf{q})$ is the body-to-world rotation matrix, $\mathbf{e}_z = [0, 0, 1]^\top$, and Newton's third law enters via $-\mathbf{F}_{rope}$ (the rope pulls the drone toward the payload).

Rotation. The rotational dynamics are described by Eq. 4:

$$\mathbf{I}\dot{\boldsymbol{\omega}} = \boldsymbol{\tau} - \boldsymbol{\omega} \times \mathbf{I}\boldsymbol{\omega}, \qquad \dot{\mathbf{q}} = \tfrac{1}{2}\,\Omega(\boldsymbol{\omega})\,\mathbf{q} \tag{4}$$

where $\mathbf{I} = \mathrm{diag}(I_x, I_y, I_z)$ and $\Omega(\boldsymbol{\omega})$ is the standard quaternion rate matrix where scalar-last quaternion convention is used here. The torque $\boldsymbol{\tau}$ is produced by the mixer matrix of the X-frame configuration: $[F_z, \tau_x, \tau_y, \tau_z]^\top = \mathbf{M}\,\mathbf{T}$. *Motor Dynamics.* Each rotor thrust T_i follows a first-order lag (PT1) seen in Eq. 5

$$\dot{T}_i = \big(T_{i,\mathrm{des}} - T_i\big)/\tau \tag{5}$$

where τ is the motor time constant and $T_{i,\mathrm{des}}$ is the output of the on-board attitude controller. *Operational Constraints.* To reflect real-platform limits, several saturation layers are applied. The total thrust vector magnitude is clipped at $F_{\max} = 82N$, derived from an approximate real-world limit considering full payload, maximum wind speed, and realistic acceleration. Individual motor thrusts are additionally saturated at $F_{\max}/4$. The horizontal velocity is bounded at $v_{h,\max} = 5\,\mathrm{m/s}$, the vertical velocity is bounded asymmetrically at $v_{v,\max} = 5\,\mathrm{m/s}$ (ascent) and $v_{v,\min} = -3\,\mathrm{m/s}$ (descent). Velocity excess above these limits generates a proportional braking correction with gain $p = 3$, applied to $\mathbf{F}_{\mathrm{des}}$ prior to the attitude controller (at the interface to the simulated drone API). The horizontal velocity is bounded at $v_{h,\max} = 5\,\mathrm{m/s}$, while the vertical velocity is bounded asymmetrically at $v_{v,\max} = 5\,\mathrm{m/s}$ (ascent) and $v_{v,\min} = -3\,\mathrm{m/s}$ (descent). Velocity excess beyond these limits generates a proportional braking correction with gain $p = 3$, applied to $\mathbf{F}_{\mathrm{des}}$ prior to the attitude controller (at the interface to the simulated drone API), since velocity constraints must be enforced at the attitude control layer interface through custom measures. The desired tilt of the body z-axis is further clipped so that roll and pitch angles cannot exceed 30°. *On-Board Attitude Controller.* The on-board attitude controller is modeled as a two-stage cascade. The outer loop maps the attitude error to a body-rate command; the inner loop maps the rate error to torque commands. The attitude error is computed geometrically from the desired thrust direction $\mathbf{z}_{\mathrm{des}} = \mathbf{F}_{\mathrm{des}}/\|\mathbf{F}_{\mathrm{des}}\|$ with Eq. 7:

$$\mathbf{e}_R = \mathbf{z}_{\mathrm{body}} \times \mathbf{z}_{\mathrm{des}}, \qquad \mathbf{e}_\omega = \boldsymbol{\omega}_{\mathrm{cmd}} - \boldsymbol{\omega}. \tag{6}$$

$$u_j = K_{p,j}\, e_j + K_{i,j} \int e_j\, \mathrm{d}t + K_{d,j}\, \frac{1 + T_{1,j}s}{1 + T_{2,j}s}\, \dot{e}_j. \tag{7}$$

where $e_j = e_{R,j}$ and $u_j = \omega_{\text{cmd},j}$ for the outer loop, and $e_j = e_{\omega,j}$ and $u_j = \tau_{\text{des},j}$ for the inner loop (see Eq. 6). Each degree of freedom can be selectively disabled by setting the corresponding gain or time constant to zero. According to the specification of the example drone, the rate command $\boldsymbol{\omega}_{cmd}$ is clamped componentwise to $[\pm 300°/\text{s}, \pm 300°/\text{s}, \pm 150°/\text{s}]$ (inner loop) and $\boldsymbol{R}_{cmd}$ to $[\pm 30°, \pm 30°, \pm\infty]$ (outer loop) *Rotor Geometry.* Physical rotors may be tilted inward by an angle α (inward-tilt) and rotated about their arm axis by an angle β (axial rotation). In the X-frame with alternating β (diagonal pairs {FL, RR} at $+\beta$, {FR, RL} at $-\beta$), the mixer matrix $\mathbf{M}$ is given in Eq. 8.

$$\begin{bmatrix} F_{z,\text{body}} \\ \tau_x \\ \tau_y \\ \tau_z \end{bmatrix} = \mathbf{M}\,\mathbf{T}, \quad \mathbf{M} = \begin{bmatrix} c_{\alpha\beta} & c_{\alpha\beta} & c_{\alpha\beta} & c_{\alpha\beta} \\ -L_r & L_r & L_r & -L_r \\ L_p & L_p & -L_p & -L_p \\ c_\psi & -c_\psi & c_\psi & -c_\psi \end{bmatrix}, \tag{8}$$

where $c_{\alpha\beta} = \cos\alpha\cos\beta$ (Thrust), $L_r = L\cos\alpha\cos\beta - c\sin\beta/\sqrt{2}$ (Roll), $L_p = L\cos\alpha\cos\beta + c\sin\beta/\sqrt{2}$ (Pitch), and $c_\psi = c\cos\beta + L\sin\beta$ (Yaw). The alternating sign of β cancels any constant yaw bias while enhancing the yaw lever arm by the term $L\sin\beta$. Small translational drift forces arising from asymmetric rotor loading under tilted rotors ($\alpha > 0$) are neglected in the mixer model, as they would require extensions to the translational dynamics but are small at the tilt angles typically encountered and cancel over time since the system operates around a hover equilibrium at 0°.

2.3 System Coupling and Integration

The rope force $\mathbf{F}_{\text{rope}}$ is computed at each integration step and applied to both ordinary differential equations (ODE), coupling the 23-dimensional system. The state is propagated using a fixed-step fourth-order Runge–Kutta integrator with $\Delta t = 0.01\,\text{s}$ in stable operating conditions. In cases where the system exhibits unstable behavior, an adaptive time-step integration scheme is employed.

2.4 Simulation Experiments

Three numerical experiments are conducted to characterize the coupled system. All experiments employ an idealized drone model with operational limits (feed-forward force compensation, no motor inertia, used thrust and velocity saturation). In addition, a realistically dynamic drone model calibrated to a DJI Matrice 210, as presented in Sect. 2.2, is used for comparison. **Simple pendulum:** During the initial swing, the drone's position is maintained at the same point through feedforward and proportional velocity control, and the pendulum moves passively due to its initial potential energy. **Frequency Response:** The transfer function from drone to pendulum XY velocity is identified via stepped-sine excitation. The drone is initialized at the steady-state deflection angle at $v_d = v_p = +v_c$ and follows a cyclic velocity profile corresponding to Eq. 9:

$$\dot{p}_d(t) = (v_c + A_{vel}\cos(2\pi f_k t))\,\hat{\mathbf{e}}, \tag{9}$$

The frequency response $H(jf_k)$ is obtained from the ratio of the projected velocity amplitudes using a single-frequency DFT (Goertzel [4]). The reference transfer function assumes an undamped ideal pendulum ($\zeta = 0$) driven by an ideal actor described in Eq. 10

$$H_{th}(j\omega) = \frac{\omega_0^2}{\omega_0^2 - \omega^2}, \quad \omega_0 = \sqrt{g/l}. \tag{10}$$

Settling Time: Settling time is evaluated via a velocity-reversal maneuver from $+v_c$ to $-v_c$ along $\hat{\mathbf{e}}$ using a constant-acceleration ramp. The pendulum is initialized at the corresponding steady-state equilibrium for $v_d = v_p = +v_c$. Settling is defined as the time until the pendulum velocity along $\hat{\mathbf{e}}$ remains within $\pm 5\%$ of the final value $-v_c$. The ramp acceleration is varied between $0.01\,g$ and $1.9\,g$ to assess its influence on settling time. In addition, the immediate step response is analyzed, as well as the additional travel distance of the pendulum relative to its initial position. The experiment is intended to characterize the system behavior in collision-avoidance scenarios.

3 Results

3.1 Mixer Sign Identification

The sign structure of the mixer matrix $\mathbf{M}$ (Eq. 8) is identified from flight data, shown in Fig. 1.

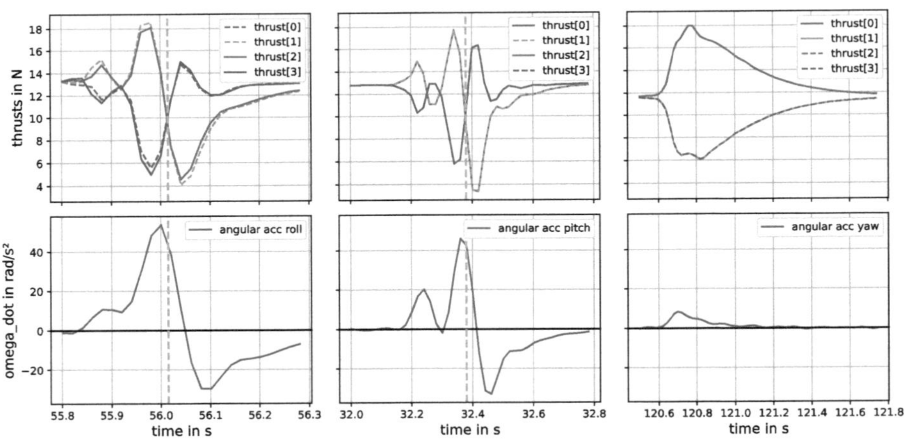

Fig. 1. Mixer sign identification for roll, pitch, and yaw maneuver. Top row: rotor thrusts; middle: angular acceleration

During dedicated roll, pitch, and yaw maneuvers, rotor thrust variations are correlated with measured angular accelerations. The resulting sign pattern is obtained by assigning $+1$ to rotors that consistently produce a positive contribution to the respective axis acceleration and -1 otherwise.

3.2 PWM-to-Thrust Calibration

Setting the total drone thrust in relation to the sum of squared rotor PWM signals (Eq. (11)) yields an approximately linear relationship as shown in Fig. 2.

$$T = a \sum_i \mathrm{PWM}_i^2 + b. \tag{11}$$

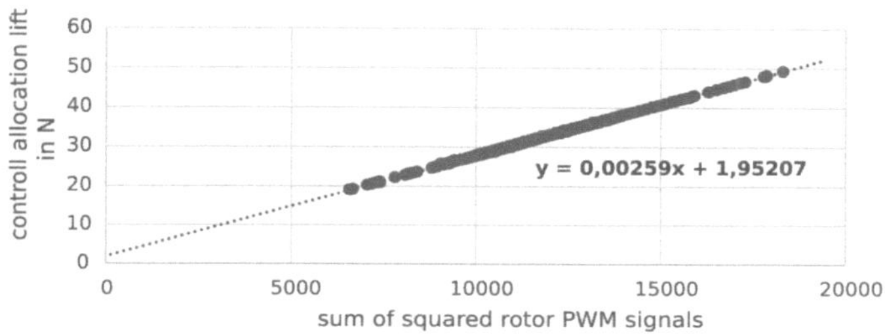

Fig. 2. Linear regression of total body-z thrust versus sum of squared rotor PWM signals, models the thrust behavior.

Using the fitted coefficients that describe the overall thrust behavior of the drone, individual rotor thrust commands are computed with the same quadratic gain but with the offset distributed equally among all rotors according to $\frac{1}{n_{\mathrm{rotors}}}$. These expressions are used to identify a subsystem of the overall system model.

3.3 Parameter Identification Results

Driving the detailed drone simulation with logged inputs and measurements validates the parameterization. A Bode analysis on the $\dot{\boldsymbol{\omega}}$ residual in Fig. 4 and timeseries of the orientation after full closed loop simulation against log data in Fig. 3 shows that the model parameters are captured accurately and confirms that the model captures the relevant frequency content.

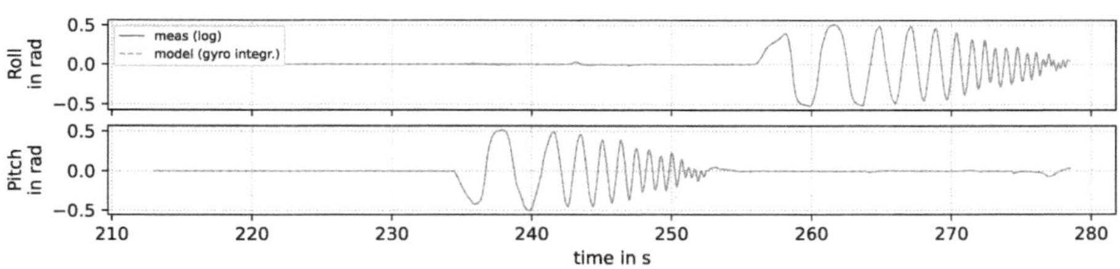

Fig. 3. Time-domain validation of the closed-loop simulation against logged flight data

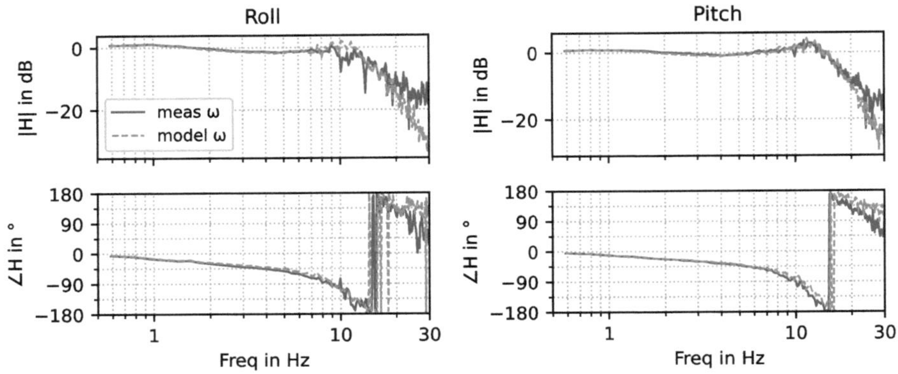

Fig. 4. Transfer function of angular rates: actual data vs. model

Figure 5 illustrates the time-series response to an initial pendulum deflection under a pre-control strategy combined with proportional control ($K_p = 5$). The simulation exhibits a destabilizing interaction between thrust and cable tension, causing the drone to drift from its initial position due to delayed actuator response. The controller regulates the target velocity in the direction of the position error, subject to saturation at a maximum of 5 m/s, while the error magnitude decreases to zero within less than 5 s as the closed-loop system compensates for deviations. The drone maintains its position with sufficient accuracy, while the pendulum swings out realistically, with rope tension exceeding gravitational force at the reversal points and amplitude decreasing as expected for a damped pendulum. A stable system behavior can be observed. The dynamics are considered physically plausible and provide a solid foundation for the development of future control strategies.

3.4 Pendulum Settling Time

This section evaluates how pendulum settling time and additional swing amplitude depend on the drone's ramp-acceleration during a velocity-reversal maneuver, to gain insights for the development of collision-avoidance systems. Figure 6 shows that the velocity settling time converges to a specific value determined by air resistance and dependent on a critical acceleration value. At higher accelerations, settling time increases significantly due to intermittent loss of rope tension during rapid counter-acceleration, with the onset threshold rising with increasing c_v. The total settling time, including the ramp phase, converges towards the pure settling time as acceleration increases, since the relative contribution of the ramp duration decreases. The maximum pendulum displacement in the direction of the initial motion decreases with increasing acceleration, even at high accelerations for which the settling time is not reached within the 40 s simulation window. Emergency braking maneuvers therefore remain below 10 m pendulum displacement for realistic accelerations on the order of $1g$.

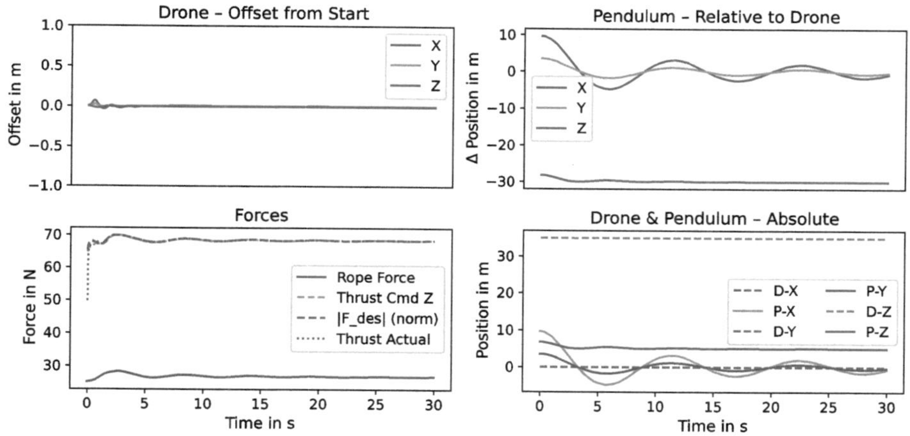

Fig. 5. Fully calibrated drone system with pre-control and proportional position-hold controller in combination with a passively reacting pendulum.

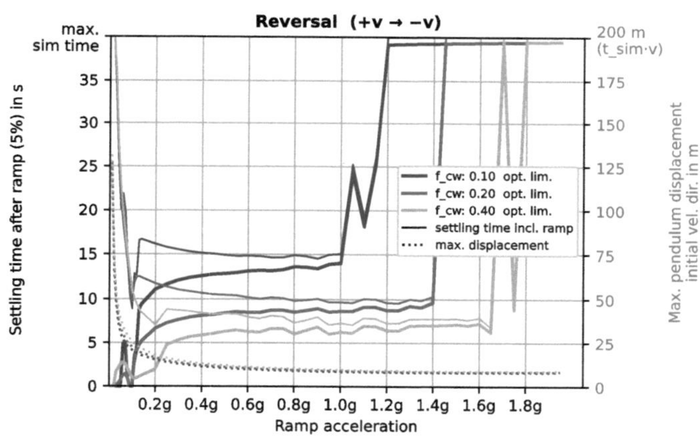

Fig. 6. Settling time and maximum pendulum displacement as a function of ramp acceleration for velocity-reversal maneuver.

3.5 Frequency Response of the Coupled System

The transfer function from drone XY velocity to pendulum XY velocity is identified via stepped-sine excitation in simulation. As shown in Fig. 7, both the idealized drone model and the dynamically realistic model exhibit similar low-frequency magnitude responses, but a magnitude offset is observed between the optimally limited model and the realistically dynamic model, which can be directly attributed to the hard velocity clipping mechanism in the limited model. The qualitative response is consistent with the behavior of a rigid actuator driving a pendulum load reported in [6]. When thrust saturation becomes dominant about 50 % of the time, a kink appears in the magnitude and phase response of the transfer function of both models. This behavior is explained by the fact that the drone can no longer achieve the commanded acceleration, effectively reducing its velocity amplitude. As a result, the pendulum partially catches up, leading to an increased magnitude ratio. At higher frequencies, rapid velocity reversals induce intermittent slackening of the tether, temporarily decoupling the payload from the drone. During these phases, the payload follows a ballistic trajectory, resulting in increased acceleration of the pendulum upon reattachment to the rope and causing the pendulum velocity to temporarily lead that of the drone. In this frequency range, contrary to the expectation that increased slack time would degrade payload tracking performance, an increase in amplitude can be observed, which is more pronounced in the idealized, fast-responding drone model due to hard clipping of the maximum velocity. The derived phase relationship explains this amplitude gain, as the pendulum accelerates more rapidly after reattachment to the tether, allowing it to temporarily catch up with the drone motion. In contrast, the dynamically realistic model exhibits reduced high-frequency response amplitudes, resulting in increased damping of the overall system and consequently less occurrence of rope slack time.

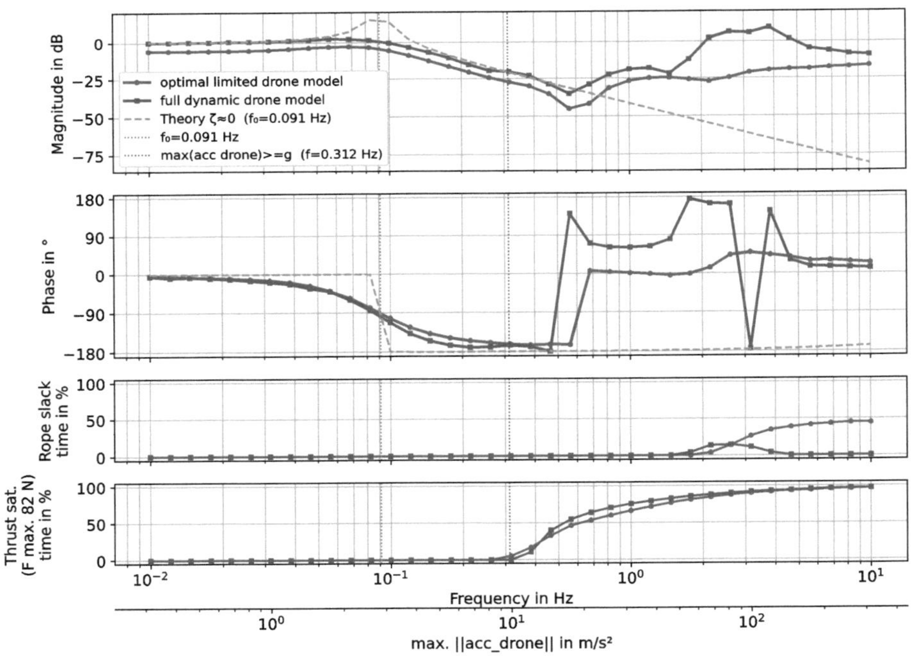

Fig. 7. Bode diagram of the drone–pendulum system ($l = 30\,\text{m}$)

4 Conclusion

The developed model, formulated as a coupled nonlinear simulation framework, captures the essential system dynamics, including rigid pendulum motion, ballistic behavior during loss of rope tension, aerodynamic effects, and full quadrotor dynamics. A key modeling contribution is the formulation of the payload as a rigid pendulum with a holonomic length constraint enforced via Lagrange multipliers. Numerical stability is ensured through Baumgarte stabilization, while the physically consistent rope force is computed without the stabilization terms, preventing artificial forces in the coupled dynamics. The drone model is calibrated up to the attitude-control level using PT2 elements and cascaded PID control loops, as well as a software-based mixer matrix enabling flexible parameterization of internal dynamics, including selective deactivation of individual PID and time-delay components. A physically motivated mixer matrix is used to determine rotational and translational accelerations. This approach allows direct specification of thrust magnitude and direction, consistent with state-of-the-art autopilot interfaces, while operating at a control level closely aligned with the physical origin of the pendulum dynamics. The resulting model reproduces key behaviors observed in simulation, including frequency-dependent coupling, tension loss phenomena, and nonlinear interaction effects between drone and payload, confirming that the proposed approach captures the dominant dynamics required for system analysis and control design. It is further observed that for lower frequency regimes up to approximately 1 Hz, no significant differences arise between the idealized, optimally responsive drone model and the dynamically realistic model, except for a small magnitude offset attributable to hard velocity clipping. The presented framework establishes a foundation for further research. Future work will focus on control design for trajectory tracking in the presence of pendulum dynamics, including fast and safe payload placement during swing motion. Furthermore, integrated collision avoidance must be addressed in conjunction with closed-loop control, simultaneously accounting for pendulum motion, drone trajectory tracking, and actuator constraints.

5 Future Work

The simulation framework is specifically suited to serve as a training and evaluation environment for learning-based control algorithms, directly addressing the workshop's focus on AI for industrial processes. In a reinforcement learning setting, the system-state forms the observation space; physically accurate ODE integration with adequate time resolution and adaptive step control ensures sim-to-real transferability. Decoupling the policy query rate from the integration step, combined with dead-time elements modeling latencies from autopilot communication and motor dynamics, allows realistic policy training. Collision-safe behavior can be promoted through reward shaping: a large terminal penalty upon actual collision, combined with a continuous proximity penalty whenever the payload approaches an obstacle below a defined safety threshold, would potentially guide the policy away from dangerous configurations while preserving the freedom to execute recovery maneuvers. For trajectory tracking under active pendulum load, data-driven or learning-based Model Predictive Control (MPC) represents a complementary direction, in which the identified system model serves as the nominal predictor and obstacle avoidance can be enforced via position constraints that are updated online as new environmental information becomes available, enabling reactive collision avoidance. A central open question is the required prediction model fidelity versus real-time feasibility. Beyond this, fundamental physical and system-theoretic limits define a performance boundary that no control strategy can overcome; characterizing these limits as a function of system parameters—such as rope length, which may be varied in flight to actively influence pendulum dynamics and meet response-time objectives—remains an open research question. The results and analyses presented in this work provide a quantitative foundation for this investigation.

Acknowledgement. This work was carried out at Hochschule Kempten – University of Applied Sciences within the third-party funded project *DraAuf*, supported under the funding call "KI-Anwendungen für den Natürlichen Klimaschutz" by Zukunft – Umwelt – Gesellschaft (ZUG) gGmbH. The authors used Claude (Anthropic) as an AI-assisted tool during the preparation of this work. Claude was used to support the generation of code components that were subsequently reviewed, tested, and validated by the authors, and for language assistance to improve the clarity and readability of the manuscript. All AI-generated content has been critically evaluated by the authors, who take full responsibility for the accuracy, integrity, and originality of this work.

Disclosure of Interests. The authors have no competing interests to declare.

References

1. Baumgarte, J.: Stabilization of constraints and integrals of motion in dynamical systems. Comput. Methods Appl. Mech. Eng. **1**(1), 1–16 (1972). https://doi.org/10.1016/0045-7825(72)90018-7
2. EHang. EH216-L Heavy-Lift Aerial Vehicle, n.d. https://www.ehang.com/ehang216l/
3. Fattori, F., Cocuzza, S.: Tethered drones: a comprehensive review of technologies, challenges, and applications. Drones **9**(6) (2025). https://doi.org/10.3390/drones9060425
4. Goertzel, G.: An algorithm for the evaluation of finite trigonometric series. Am. Math. Mon. **65**(1), 34–35 (1958)
5. Mahony, R., Kumar, V., Corke, P.: Multirotor aerial vehicles: modeling, estimation, and control of quadrotor. IEEE Robot. Autom. Mag. **19**(3), 20–32 (2012). https://doi.org/10.1109/MRA.2012.2206474
6. Nishiyama, Y., Ishikawa, J.: Crane operation support system with damping control function using model predictive control. In: 2025 IEEE International Conference on Industrial Technology (ICIT), pp. 1–6 (2025). https://doi.org/10.1109/ICIT63637.2025.10965219
7. Patterson, T.: Zipline receives FAA part 135 carrier certificate (2022). FLYING Magazine. https://www.flyingmag.com/zipline-receives-faa-part-135-carrier-certificate/

8. Sreenath, K., Lee, T., Kumar, V.: Geometric control and differential flatness of a quadrotor UAV with a cable-suspended load. In: 52nd IEEE Conference on Decision and Control, pp. 2269–2274 (2013). https://doi.org/10.1109/CDC.2013.6760219
9. SZ DJI Technology Co., Ltd. DJI FlyCart 100 (2026). https://www.dji.com/de/flycart-100
10. Voigt, J.-L., Happich, G., Weiser, T.: Development and validation of a rotary impact drilling system for an ultra-lightweight UAV-deployed tree planting robot enabling autonomous forest restoration. In: Eurocast 2026: Extended Abstracts, Las Palmas de Gran Canaria, Spain (2026). Extended Abstracts, February 23–27, Hochschule Kempten

Federated Learning for Multivariate Time Series Anomaly Detection in Industrial Automation

Khayyam Nosrati[1(✉)], Martin Uray[1,2], Saverio Messineo[1], Olaf Sassnick[1], and Stefan Huber[1]

[1] Josef Ressel Centre for Intelligent and Secure Industrial Automation, Salzburg University of Applied Sciences, Puch/Hallein, Austria
{khayyam.nosrati,martin.uray}@fh-salzburg.ac.at
[2] Department of Artificial Intelligence and Human Interfaces, Paris Lodron University of Salzburg, Salzburg, Austria

Abstract. Federated learning (FL) has broadened the horizon for multivariate time series anomaly detection (MTSAD). However, benchmarking such anomaly detection methods within FL paradigm poses data-centric challenges. The existing datasets do not counteract these challenges since they do not simultaneously provide sufficient scale, accurate labels, and freedom from common flaws. In addition, the role of cyclic process behavior, which is common in discrete industrial automation, remains underexplored for MTSAD for the current state of research. This paper aims to shed more light on the literature and address these gaps by introducing a dataset designed with cyclic dynamics arising from the repetitive nature of discrete automation processes and evaluates selected MTSAD methods on both the proposed dataset and a public benchmark dataset.

Keywords: Federated Learning · Hierarchical Federated Learning · Multivariate Time Series · Anomaly Detection

1 Introduction

Operational Technology (OT) systems [22], such as Industrial Control System (ICS), are distributed, embedded cyber-physical infrastructures designed for real-time monitoring and control of industrial processes, and are the backbone of critical industrial sectors. The advent of intelligent technologies has enhanced the capabilities of these systems to better leverage data and foster a new era of automated and intelligent computing. One of the most prominent applications of AI in OT environments is anomaly detection; and more particularly, anomaly detection on Multivariate Time Series (MTS) data acquired through the observation of dynamic processes.

Recent research has introduced advanced Deep Learning (DL) techniques for anomaly detection, demonstrating high potential. DL scales with input data

© The Author(s), under exclusive license to Springer Nature Switzerland AG 2027
M. Geiß et al. (Eds.): DEXA 2026 Workshops, CCIS 3078, pp. 90–101, 2027.
https://doi.org/10.1007/978-3-032-34202-7_8

volume in various tasks [5,19], which is one of the core motivation to centralizing data and learning an effective machine learning model on a central instance. This paradigm is known as Centralized Learning (CL) which typically handles massive computation and enormous storage demands on cloud servers.

Alternatively, Federated Learning, and Hierarchical Federated Learning (H-FL) provide distributed computation and the necessary communication topology to address the challenges of risen by CL.

In this work, we investigate the application of MTSAD in an unsupervised setting within the paradigms of CL, FL, and H-FL. We evaluate performance trade-offs on datasets with *different* characters, while we primarily focus on horizontal federated learning and centralized learning setups. In horizontal FL local datasets at edge nodes have identical and shared feature space, while sample space varies. Therefore, this study assumes homogeneous neural network architectures across edge nodes.

Contributions. In this work, we identified the following key research gaps: (i) lack of comparative studies on federated and hierarchical federated MTSAD, (ii) absence of FL repetitive MTSAD-specific datasets out of industrial production cycles. In addition, the lack of a unified definition of industrial anomaly, combined with the use of multiple, sometimes flawed evaluation metrics, prevents the development of consistent and robust assessment practices to advance industry and automation. To address these gaps, we make the following contributions:

- Introduce a new suite of ten repetitive multivariate time series anomaly detection datasets.
- Provide comparative insights on the five MTSAD methods that curated by a principled approach—based on problem-solving specificity, architectural suitability for MTSAD, and integrability to available FL software frameworks while conserving diversity of architectures—across centralized, federated, and hierarchical-federated learning paradigms on repetitive and non-repetitive datasets.

2 Related Work

Multivariate Time Series Anomaly Detection has long been approached in localized learning environments, where machine learning is performed on a single device using a single dataset. Existing MTSAD approaches can broadly be categorized into statistical, forecasting-based [30], and reconstruction-based [26] methods. More recently, MTSAD is approached by federated learning such as FATRAF [25] which proposes a lightweight architecture designed for communication efficiency investigated in industrial environments, while [28] are studied in non-industrial applications. Different aggregation strategies are utilized by such methods: FATRAF and [7] adapt FedAvg, FedUAD [32] employs the FedCC, and [28] integrates FLTrELM. Other advancements involve asynchronous communication protocol [18]. Hierarchical federated learning has also been considered for MTSAD, as demonstrated by HFL-ADS [2] and HFed-IDS [24], which apply hierarchical variants of FedAvg incorporating intermediate aggregation levels. Despite these developments, benchmarking studies comparing federated and hierarchical federated learning approaches for MTSAD remain largely absent.

Several highly cited public benchmarks are repeatedly used for evaluating MTSAD, either as single datasets or as collections of similar datasets [1,6,8,23]. Server Machine Dataset (SMD) is a collection of 28 datasets which represent a 5-week-long data collected from a large internet company. Exathlon is constituted by recording the repeated executions of 10 different Spark streaming applications on a 4-node cluster. NASA's Mars Science Laboratory (MSL) and Soil Moisture Active Passive (SMAP) datasets contain multivariate time series where one variable representing a sensor measurement and the remaining variables represent one-hot encoded telemetry commands. The observations in SMD, MSL, SMAP are all equally-spaced one minute apart, while in Exathlon are one second apart. Additional benchmarks such as Secure Water Treatment (SWaT), Water Distribution (WaDi), and Pooled Server Metrics (PSM) are commonly employed as single datasets for MTSAD. In federated settings, these datasets are typically adapted through partitioning strategies, where subsets of data are assigned to multiple edge nodes. However, prior analyses have highlighted that these suffer from inherent limitations or flaws [10,27,29]. Besides, these datasets lack explicit cyclic properties.

In conclusion, there is scarcity of datasets suitable for MTSAD that capture discrete, repetitive, and state-driven characteristics of industrial production cycles, including pick-and-place operations, injection molding, assembly, and in-line component testing. Furthermore, current literature rarely investigates these scenarios through joint benchmarking of federated and hierarchical federated learning approaches. These shortcomings collectively motivate the focus of this work.

3 Preliminaries

3.1 Federated Learning

In this work, centralized, federated, and hierarchical federated learning settings follow a centralized communication topology. The choice of FL aggregation strategy is *FedAvg* [13] for its simplicity of implementation, interpretability and strong performance reported in the literature. For H-FL aggregation, HierFAVG [11]—an extended version of FedAvg that has theoretical convergence guarantees—is adopted. The default optimizers of the chosen methods are set to SGD for FL and H-FL simulations, because FedAvg and HierFAVG use SGD as their local optimizers.

3.2 Evaluation Metrics

This work adopts a centralized evaluation protocol that considers the union of all available windowed test datasets and uses it at each communication round or epoch to evaluate candidate model performance. Four evaluation metrics were selected from available MTSAD metrics [12,21] to ensure application-agnostic performance assessment. Specifically, the threshold-based metrics are point-wise F1-score and composite F1-score [4], while the non-threshold-based metrics are Area Under the Precision-Recall Curve (AUC-PR) and Volume Under the PrecisionRecall Surface (VUS-PR) introduced by [17].

3.3 Datasets

In this work, we use Application Server Dataset (ASD) proposed by [10] as a benchmark that comprises 12 datasets designed for multivariate time series anomaly detection and interpretation. They are collected from a large internet company, each of the entities has 19 metrics characterizing status of the server. The time series constitutes 19 variables and data points that are equally spaced for every 5 min. The overall anomaly ratio is 4.61%. ASD is chosen for avoiding the following challenges: (i) Information Sparsity or Loss, (ii) Concept Drift, (iii) Mixed-valued Variables.

In this paper, we propose Quanser Aero 2 Pick-and-Place Datase (QAPPD), a new collection of publicly available multivariate time series data [16]. It is explicitly made for MTSAD in federated learning. QAPPD serves as the second dataset used for experimental evaluation in this study. It consists of ten independent train–test pairs, each of which can be interpreted as data local to federated learning edge nodes. It offers a physically grounded and semantically interpretable data-generation process. The dynamics were inspired by robotic pick and place operations, typically comprise multiple mechanical components—rotary joints, arms, and a gripper (mechanical hand) to pick and place items, carousels to transport these items, and visual inspection for verifying the existence the items and communicating their position. Along with all the structural components mentioned, the design considered nuances such as elevation of the platforms, composition of the components, distances to one another, and the proportions of these parts. To control the degree of statistical heterogeneity across datasets, the number of system components and the number of trajectory waypoints were kept constant across all ten train–test pairs, i.e. each configuration contains one articulated robot and two carousel. The resulting system behavior is characterized by cyclic trajectories in the system's state space, where the gripper follows a smooth, periodic motion pattern resembling a cycloid-like motion. The trajectories were generated using the tool provided by [20].

To instantiate the above process, a cyber-physical system (CPS) called Quanser Aero 2 was used, i.e. a fully integrated dual-motor experimental setup, designed for advanced control research for aerospace and aeronautics applications. This device has two degrees of freedom (DOF), which are used to model a simplified robotic arm in this work. In this abstraction, the front fan is treated as the robotic gripper, while the body of the device represents a planar 2-DOF arm. The dataset variables are listed in Table 1.

Dataset Design Principles. The dataset was constructed using design semantics that explicitly encode normal and anomalous behavior. This approach significantly reduces the manual labeling effort, as anomalies are known by design.

Properties of the Dataset. The multivariate time series represents the system trajectory, with the observed variables jointly defining the system state at each discrete time step. In addition, the time series exhibits approximately periodic behavior.

Table 1. Detailed CPS data on key observed variables.

Name	Description	Unit	Range
voltage0	DC-motor 0 voltage	V	± 18
voltage1	DC-motor 1 voltage	V	± 18
current0	DC-motor 0 current	A	0.540
current1	DC-motor 1 current	A	0.540
motorSpd0	Rotational speed of fan	rpm	3050
motorSpd1	Rotational speed of fan	rpm	3050
pitch	Actual pitch angle	rad	$\pm 2\pi/9$
yaw	Actual yaw angle	rad	2π
pitchDot	Pitch's angular velocity	rad s^{-1}	—
yawDot	Yaw's angular velocity	rad s^{-1}	—

Anomaly Design from Two Sources

- Anomalies induced by system dynamics:
  - Let $D = [0.3, 0.7]$ denote the set representing the expected range of pause or delay duration (in seconds) at pick-up anticipation positions. Then, any value falling outside D is deemed anomalous.
  - Let $P = [3, 5]$ be the set representing normal pick-up angular positions (in degrees). We consider any value not in P anomalous.
- Anomalies induced by the system controller:
  - Let $\bar{V} = [-18, 18]$ be the set representing the nominal voltage range for DC motors. Then any value outside $\bar{V}$ is considered anomalous.

Data Acquisition. The dataset was originally recorded at 500 Hz and downsampled to 50 Hz to reduce computational overhead, resulting in equally distant time steps of 0.02 s. Figure 1 illustrates an example anomaly in the test set.

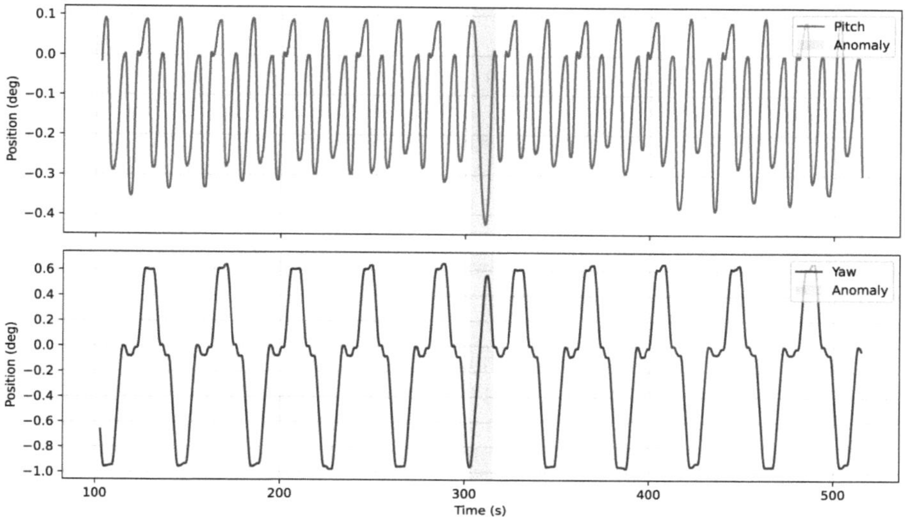

Fig. 1. This snippet captures around 400 s of recorded data. Red highlights a misposition anomaly at a picking position affecting both yaw and pitch variables.

3.4 Baselines and Architectures

Five representative Multivariate Time Series Anomaly Detection (MTSAD) baseline methods are selected based on their compatibility with the employed FL software frameworks and their architectural suitability. In particular, methods requiring a single optimization phase are preferred, excluding approaches with additional optimization during evaluation (e.g., MAD-GAN [9]). The selection further considers inference latency via forward-pass time complexity, problem scope, and feature extraction capabilities, specifically the ability to model temporal (intra-variable) and spatial (inter-variable) dependencies. The curated methods and their characteristics are summarized in Table 2.

Table 2. Summarizing how suitable the MTSAD methods are in terms of the aforementioned factors. Time complexity is per dominant layer(s) rather than total. m: sequence length. d: variable dimension. z: latent dimension. k: kernel size. $d = 1$ for 1D convolution but for MTSAD, equals to the number of variables m. UTSAD: Univariate Time Series Anomaly Detection.

Method	Complexity	Problem	Temporal	Spatial
USAD [3]	$O(m \cdot d \cdot z)$	MTSAD	✗	✗
DeepAnT [15]	$O(m \cdot d \cdot k \cdot z)$	UTSAD	✓	✗
LSTM-AE [14]	$O(m \cdot d \cdot z^2)$	MTSAD	✓	✗
TranAD [26]	$O(m^2 \cdot d \cdot z)$	MTSAD	✓	✓
MTAD-GAT [31]	$O(m^2 \cdot d \cdot z + m \cdot d^2 \cdot z)$	MTSAD	✓	✓

4 Experimental Setup

Software. The computer cluster used to run experiments of this study, runs on Ubuntu 20.04 LTS (Focal Fossa) and is managed for runs by SLURM Workload Manager. Python programming language was chosen for its versatility and the core implementations of this paper are done with it. Lightning is a high level framework for PyTorch, was selected to keep the unnecessary engineering code abstracted away, allowing us to focus on the problem-solving. Flower is chosen for FL simulations, and a minimal framework is self-developed for H-FL simulations utilizing GPU for parallel processing.

Hyperparameter Tuning. To ensure fair comparisons, the algorithms are configured using their best searched hyperparameters preceding the main experiment runs. 70% of test and 80% of train sets were sampled to tune the hyperparameters, speeding up the hyperparameter search as well as ensuring the representativeness of the sets. The objective for the search is based on maximizing point-wise F1-score. The batch size is set to 1024, however, for the trainings that are run with distributed training strategy, the batch size was divided by the number of GPUs to obtain the same results as if they are run on single GPU. The hyperparameter sweep were carried out using GPU's Tensor cores that support reduced precision format TensorFloat-32 (**TF32**), while using CUDA cores that support standard single-precision floating-point (**FP32**) for the main experiments. TF32 allows for using the same dynamic range as FP32 (8-bit exponent) but lower precision, i.e. 10-bit mantissa, compared to 23 bits. This is done to trade off some numerical accuracy with speed and memory for the heavy task of hyperparameter search.

Hardware. The experiments are performed on a single computer cluster node that includ6es 4x NVIDIA RTX A6000 graphic card with 48 GB vRAM each, 2x AMD EPYC 7452 with 2.35 GHz frequency with total of 64 computing cores.

Main Experiments. To obtain meaningful results, the experiments[1] are independently run five times with different initialization seeds. The corresponding results for each unique experiment is reported as the mean plus-minus one standard deviation. The results are reported from the best fit state of the models.

5 Results

Detection performances are reported in Table 3. Under centralized learning, USAD, LSTM-AE, and DeepAnT outperform TranAD and MTAD-GAT overall. LSTM-AE shows stable performance across both datasets, USAD performs best on QAPPD, and DeepAnT on ASD, despite its weaker performance on QAPPD.

[1] The code for the experiments are publicly available https://github.com/JRC-ISIA/industrial-federated-learning/.

Table 3. Detection performance across different setups, reported as mean ± standard deviation over five independent runs. F1 denotes the standard point-wise F1 score; $F1_c$ denotes the composite F1 score.

Setting	Dataset	Method	F1↑	$F1_c$↑	VUS-PR↑	AUC-PR↑
Centralized	ASD	USAD	57.247 ± 0.122	61.369 ± 0.082	45.676 ± 3.430	42.020 ± 2.712
		DeepAnT	57.025 ± 0.000	**84.166 ± 0.459**	55.976 ± 0.725	51.147 ± 0.656
		LSTM-AE	**63.456 ± 0.963**	82.117 ± 3.394	**72.900 ± 2.027**	**72.615 ± 1.768**
		TranAD	57.032 ± 0.009	60.108 ± 2.362	44.077 ± 0.493	40.110 ± 0.129
		MTAD-GAT	57.033 ± 0.011	66.179 ± 6.240	45.942 ± 3.587	41.216 ± 2.801
	QAPPD	USAD	**23.849 ± 0.891**	37.035 ± 2.728	13.030 ± 0.494	19.777 ± 0.768
		DeepAnT	11.711 ± 2.849	18.338 ± 4.439	6.613 ± 0.885	6.110 ± 0.474
		LSTM-AE	28.293 ± 8.387	**50.181 ± 10.358**	**17.238 ± 5.619**	**23.365 ± 6.600**
		TranAD	13.910 ± 2.233	18.706 ± 4.938	7.907 ± 1.056	7.159 ± 1.369
		MTAD-GAT	13.284 ± 1.547	19.172 ± 4.908	8.082 ± 0.591	7.425 ± 1.349
Federated	ASD	USAD	57.025 ± 0.000	57.398 ± 0.005	43.325 ± 0.007	41.044 ± 0.004
		DeepAnT	57.025 ± 0.000	**85.382 ± 0.874**	**55.379 ± 0.318**	**50.664 ± 0.321**
		LSTM-AE	57.025 ± 0.000	57.2785 ± 0.162	43.544 ± 0.327	41.407 ± 0.475
		TranAD	**57.062 ± 0.034**	59.003 ± 1.897	44.2354 ± 0.381	40.076 ± 0.285
		MTAD-GAT	57.061 ± 0.039	62.553 ± 1.935	46.442 ± 1.183	41.859 ± 1.186
	QAPPD	USAD	**33.230 ± 0.155**	**56.412 ± 0.398**	**19.137 ± 0.109**	**26.651 ± 0.054**
		DeepAnT	13.0456 ± 0.375	15.421 ± 0.586	7.735 ± 0.076	5.661 ± 0.157
		LSTM-AE	25.139 ± 7.059	35.909 ± 14.888	13.872 ± 4.375	18.773 ± 5.920
		TranAD	12.486 ± 1.667	16.195 ± 3.115	6.503 ± 1.047	5.905 ± 0.867
		MTAD-GAT	12.097 ± 1.110	19.671 ± 3.481	7.221 ± 0.424	7.009 ± 0.715
Hierarchical Federated	ASD	USAD	57.023 ± 0.000	57.408 ± 0.005	40.975 ± 0.006	41.060 ± 0.005
		DeepAnT	57.024 ± 0.003	64.718 ± 2.681	**41.860 ± 0.310**	41.039 ± 0.489
		LSTM-AE	57.023 ± 0.000	57.189 ± 0.119	41.340 ± 0.479	41.295 ± 0.577
		TranAD	**57.024 ± 0.002**	60.423 ± 1.501	41.323 ± 0.208	40.079 ± 0.237
		MTAD-GAT	57.023 ± 0.008	**78.937 ± 2.255**	**44.705 ± 0.608**	**43.686 ± 0.528**
	QAPPD	USAD	**34.612 ± 0.058**	65.767 ± 0.141	**22.564 ± 0.024**	**30.053 ± 0.045**
		DeepAnT	11.631 ± 0.252	13.468 ± 0.444	5.275 ± 0.194	5.212 ± 0.298
		LSTM-AE	26.337 ± 7.008	47.693 ± 10.960	15.710 ± 4.754	21.476 ± 5.929
		TranAD	12.022 ± 2.624	18.982 ± 7.489	6.236 ± 0.967	6.120 ± 1.483
		MTAD-GAT	11.855 ± 1.322	33.084 ± 2.783	7.356 ± 0.620	8.820 ± 0.764

MTAD-GAT and TranAD generally struggle in this setting. In federated learning, USAD and LSTM-AE achieve satisfactory results on QAPPD but remain less competitive on ASD, whereas DeepAnT and MTAD-GAT perform better on ASD. TranAD is consistently the weakest method. In hierarchical federated learning, USAD and LSTM-AE perform best on QAPPD, MTAD-GAT and DeepAnT excel on ASD, and TrandAD falls short on QAPPD while moderately succeeds on ASD.

To quantify performance retention across the two datasets and three learning paradigms, the values of corresponding metrics in FL and H-FL were divided by those of CL and shown in Fig. 2. Each cell shows the ratio of a given metric value obtained under FL or H-FL to that of CL, illustrating the sensitivity of each method to the distributed learning paradigm and dataset characteristics.

Notably, federated and hierarchical federated settings, although with different degrees of distribution, did not substantially affect model behaviors. MTAD-GAT and TranAD were the most robust, showing the smallest performance drop. LSTM-AE and DeepAnT suffered the largest degradations across metrics. Generally, USAD showed an unexpected improvement compared to other baselines.

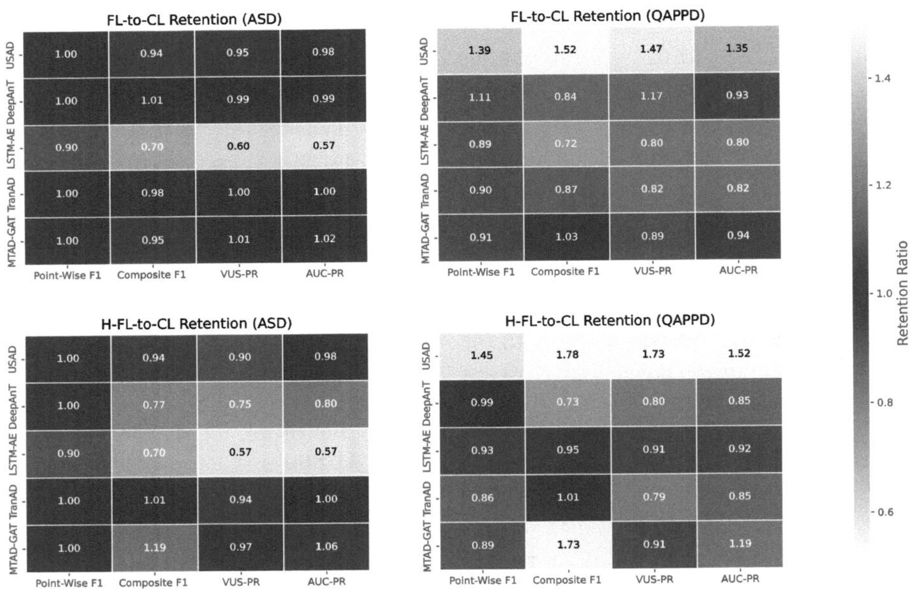

Fig. 2. Relative performance in federated and hierarchical federated learning settings compared to centralized learning, where a ratio of 1.0 denotes parity.

6 Discussion

The empirical results indicate that architectural heterogeneity and the adoption of state-of-the-art models, such as Transformers and GNNs, do not inherently lead to improved anomaly detection performance. This suggests that increased model capacity and architectural sophistication may introduce unnecessary modeling overhead, particularly in repetitive industrial time series. Instead, less complex methods consistently achieve stronger results across settings. Notably, models that either explicitly capture temporal dependencies—such as LSTM- and TCN-based architectures—or employ effective training objectives, as in USAD with fully connected layers under an adversarial loss formulation, demonstrate superior performance. This suggests that alignment between model inductive bias, training strategy, and the characteristics of repetitive industrial time series is more critical than architectural complexity alone.

7 Conclusion

The comparative evaluation conducted in this paper reveals that the effectiveness of anomaly detection methods is tightly coupled to both the learning paradigm in which they operate and the characteristics of the datasets on which they are evaluated. These findings underscore the importance of considering learning topology and data structure jointly when assessing multivariate time series anomaly detection methods. A limitation of this work is that, while a new dataset is proposed and empirical benchmarking is provided, these results are not supported by a rigorous analysis that would enable deeper insights. Looking forward, several research directions naturally emerge from this work, including extension of anomaly definitions beyond bounded deviations to incorporate multiscale notions of abnormality, and integrating domain knowledge into learning objectives and selecting the architectural priors that align with the underlying system dynamics to support explainable and trustworthy anomaly detection.

Acknowledgments. The financial support by the Christian Doppler Research Association, the Austrian Federal Ministry for Digital and Economic Affairs and the Federal State of Salzburg is gratefully acknowledged.

References

1. Abdulaal, A., Liu, Z., Lancewicki, T.: Practical approach to asynchronous multivariate time series anomaly detection and localization. In: Proceedings of the 27th ACM SIGKDD Conference on Knowledge Discovery and Data Mining, KDD 2021, pp. 2485–2494. Association for Computing Machinery, New York (2021). https://doi.org/10.1145/3447548.3467174
2. Alshede, H., Jambi, K., Nassef, L., Alowidi, N., Fadel, E.: FedAvg-P: performance-based hierarchical federated learning-based anomaly detection system aggregation strategy for advanced metering infrastructure. Sensors **24**(17) (2024). https://doi.org/10.3390/s24175492
3. Audibert, J., Michiardi, P., Guyard, F., Marti, S., Zuluaga, M.A.: USAD: unsupervised anomaly detection on multivariate time series. In: Proceedings of the 26th ACM SIGKDD International Conference on Knowledge Discovery & Data Mining, KDD 2020, pp. 3395–3404. Association for Computing Machinery, New York (2020). https://doi.org/10.1145/3394486.3403392

4. Garg, A., Zhang, W., Samaran, J., Savitha, R., Foo, C.S.: An evaluation of anomaly detection and diagnosis in multivariate time series. IEEE Trans. Neural Netw. Learn. Syst. **33**(6), 2508–2517 (2022). https://doi.org/10.1109/TNNLS.2021.3105827
5. Hestness, J., et al.: Deep learning scaling is predictable, empirically, arXiv preprint arXiv:1712.00409 (2017)
6. Hundman, K., Constantinou, V., Laporte, C., Colwell, I., Soderstrom, T.: Detecting spacecraft anomalies using LSTMs and nonparametric dynamic thresholding. In: Proceedings of the 24th ACM SIGKDD International Conference on Knowledge Discovery & Data Mining, KDD 2018, pp. 387–395. Association for Computing Machinery, New York (2018). https://doi.org/10.1145/3219819.3219845
7. Huong, T.T., et al.: Detecting cyberattacks using anomaly detection in industrial control systems: a federated learning approach. Comput. Ind. **132**, 103509 (2021). https://doi.org/10.1016/j.compind.2021.103509
8. Jacob, V., Song, F., Stiegler, A., Rad, B., Diao, Y., Tatbul, N.: Exathlon: a benchmark for explainable anomaly detection over time series. Proc. VLDB Endow. **14**(11), 2613–2626 (2021). https://doi.org/10.14778/3476249.3476307
9. Li, D., Chen, D., Jin, B., Shi, L., Goh, J., Ng, S.-K.: MAD-GAN: multivariate anomaly detection for time series data with generative adversarial networks. In: Tetko, I.V., Kůrková, V., Karpov, P., Theis, F. (eds.) ICANN 2019. LNCS, vol. 11730, pp. 703–716. Springer, Cham (2019). https://doi.org/10.1007/978-3-030-30490-4_56
10. Li, Z., et al.: Multivariate time series anomaly detection and interpretation using hierarchical inter-metric and temporal embedding. In: Proceedings of the 27th ACM SIGKDD Conference on Knowledge Discovery & Data Mining, pp. 3220–3230, KDD 2021. Association for Computing Machinery, New York (2021)
11. Liu, L., Zhang, J., Song, S., Letaief, K.B.: Client-edge-cloud hierarchical federated learning. In: ICC 2020-2020 IEEE International Conference on Communications (ICC), pp. 1–6. IEEE (2020)
12. Liu, Q., et al.: The elephant in the room: towards a reliable time-series anomaly detection benchmark. In: Zhang, C. (ed.) Advances in Neural Information Processing Systems, vol. 37, pp. 108231–108261. Curran Associates, Inc (2024)
13. McMahan, B., Moore, E., Ramage, D., Hampson, S., y Arcas, B.A.: Communication-efficient learning of deep networks from decentralized data. In: Artificial Intelligence and Statistics, pp. 1273–1282. PMLR (2017)
14. Mirza, A.H., Cosan, S.: Computer network intrusion detection using sequential LSTM neural networks autoencoders. In: 2018 26th Signal Processing and Communications Applications Conference (SIU), pp. 1–4 (2018)
15. Munir, M., Siddiqui, S.A., Dengel, A., Ahmed, S.: DeepAnT: a deep learning approach for unsupervised anomaly detection in time series. IEEE Access **7**, 1991–2005 (2019)
16. Nosrati, K., Uray, M., Messineo, S., Sassnick, O., Huber, S.: Quanser aero 2 pick-and-place dataset (QAPPD) (2026). https://doi.org/10.5281/zenodo.20287835
17. Paparrizos, J., Boniol, P., Palpanas, T., Tsay, R.S., Elmore, A., Franklin, M.J.: Volume under the surface: a new accuracy evaluation measure for time-series anomaly detection. Proc. VLDB Endow. **15**(11), 2774–2787 (2022)
18. Raeiszadeh, M., Ebrahimzadeh, A., Glitho, R.H., Eker, J., Mini, R.A.: Asynchronous real-time federated learning for anomaly detection in microservice cloud applications. IEEE Trans. Mach. Learn. Commun. Netw. **3**, 176–194 (2025)
19. Sarker, I.H.: Deep learning: a comprehensive overview on techniques, taxonomy, applications and research directions. SN Comput. Sci. **2**(6), 1–20 (2021)

20. Saßnick, O., Schäfer, G., Rosenstatter, T., Huber, S.: A generative model based honeypot for industrial OPC UA communication. In: Computer Aided Systems Theory – EUROCAST 2024, pp. 320–334. Springer, Cham (2025). https://doi.org/10.1007/978-3-031-83885-9_29
21. Sørbø, S., Ruocco, M.: Navigating the metric maze: a taxonomy of evaluation metrics for anomaly detection in time series. Data Min. Knowl. Disc. **38**(3), 1027–1068 (2024)
22. Stouffer, K., et al.: Guide to operational technology (OT) security. NIST Special Publication (SP) NIST SP 800-82r3, National Institute of Standards and Technology, Gaithersburg, MD (2023). https://doi.org/10.6028/NIST.SP.800-82r3
23. Su, Y., Zhao, Y., Niu, C., Liu, R., Sun, W., Pei, D.: Robust anomaly detection for multivariate time series through stochastic recurrent neural network. In: Proceedings of the 25th ACM SIGKDD International Conference on Knowledge Discovery & Data Mining, KDD 2019, pp. 2828–2837. Association for Computing Machinery, New York (2019). https://doi.org/10.1145/3292500.3330672
24. Sun, X., et al.: A hierarchical federated learning-based intrusion detection system for 5g smart grids. Electronics **11**(16) (2022). https://doi.org/10.3390/electronics11162627
25. Truong, H.T., et al.: Light-weight federated learning-based anomaly detection for time-series data in industrial control systems. Comput. Ind. **140**, 103692 (2022). https://doi.org/10.1016/j.compind.2022.103692
26. Tuli, S., Casale, G., Jennings, N.R.: TranAD: deep transformer networks for anomaly detection in multivariate time series data. Proc. VLDB Endow. **15**(6), 1201–1214 (2022). https://doi.org/10.14778/3514061.3514067
27. Wagner, D., Michels, T., Schulz, F.C.F., Nair, A., Rudolph, M.R., Kloft, M.: TimesEAD: benchmarking deep multivariate time-series anomaly detection. Trans. Mach. Learn. Res. (2023)
28. Wang, K., Li, J., Wu, W.: An efficient intrusion detection method based on federated transfer learning and an extreme learning machine with privacy preservation. Secur. Commun. Netw. **2022**(1), 2913293 (2022). https://doi.org/10.1155/2022/2913293
29. Wu, R., Keogh, E.J.: Current time series anomaly detection benchmarks are flawed and are creating the illusion of progress (extended abstract). In: 2022 IEEE 38th International Conference on Data Engineering (ICDE), pp. 1479–1480 (2022)
30. Xu, J., Wu, H., Wang, J., Long, M.: Anomaly transformer: time series anomaly detection with association discrepancy, arXiv preprint arXiv:2110.02642 (2021)
31. Zhao, H., et al.: Multivariate time-series anomaly detection via graph attention network. In: 2020 IEEE International Conference on Data Mining (ICDM), pp. 841–850. IEEE (2020)
32. Zhu, W., et al.: Deep federated anomaly detection for multivariate time series data. In: 2022 IEEE International Conference on Big Data (Big Data), pp. 1–10 (2022). https://doi.org/10.1109/BigData55660.2022.10064694

Anticipatory Reinforcement Learning for Trajectory Tracking

Georg Schäfer[1,2(✉)], Jakob Rehrl[1], Stefan Huber[1], and Simon Hirlaender[2]

[1] Josef Ressel Centre for Intelligent and Secure Industrial Automation, Salzburg University of Applied Sciences, Salzburg, Austria
[2] Paris Lodron University of Salzburg, Salzburg, Austria
georg.schaefer@fh-salzburg.ac.at

Abstract. Deep reinforcement learning (DRL) in industrial control often suffers from lag and overshoot due to purely reactive control based on the current tracking error. To achieve anticipatory control without high computational overhead, we introduce a predictive formulation that augments the DRL state space with target velocities and future reference horizons. Evaluating eight configurations using proximal policy optimization (PPO) on a 1-degree-of-freedom (1-DoF) helicopter testbed, simulation results showed a 9-fold error reduction, lowering the mean absolute deviation from 2.73° to 0.31°. However, zero-shot transfer to physical hardware revealed a sim-to-real gap. Interestingly, a simpler configuration using a single, further look-ahead horizon matched the real-world top performance of the most complex model (1.11°). Overall, evaluating various combinations of prediction horizons and target velocities demonstrated that highly granular predictive data is not necessarily required for physical transfer.

Keywords: Predictive Reinforcement Learning · Anticipatory Control · Trajectory Tracking

1 Introduction

Industrial cyber-physical systems (ICPSs) integrate physical processes with computational control and underpin modern manufacturing, energy management, and smart production lines. Reinforcement learning (RL) has demonstrated success in optimizing these complex, dynamic environments, offering data-driven control solutions where mathematical modeling is challenging [2]. However, many deep reinforcement learning (DRL) policies typically compute their actions based on the current state and the instantaneous reference, without anticipating future changes in the reference trajectory. For example, standard benchmark environments and baseline controllers often rely solely on the instantaneous deviation from a target (e.g., [1,4]). Consequently, these agents frequently exhibit lag and overshoot, as they cannot anticipate upcoming trajectory changes. In contrast, classic control strategies such as model predictive control (MPC) systematically

© The Author(s), under exclusive license to Springer Nature Switzerland AG 2027
M. Geiß et al. (Eds.): DEXA 2026 Workshops, CCIS 3078, pp. 102–106, 2027.
https://doi.org/10.1007/978-3-032-34202-7_9

incorporate future reference trajectories. However, computing optimal actuating sequences requires solving an optimization problem at every time step. This high computational demand limits MPC in real-time industrial applications. DRLoffers computationally lightweight online inference once trained. To bridge the gap between the foresight of MPC and the computational efficiency of DRL, this work-in-progress paper introduces a predictive problem formulation for industrial mechatronic systems. Our primary contributions are twofold: (i) We embed "foresight" into the DRL state space by augmenting the agent's observation with future target variables and the current target velocity. (ii) We systematically evaluate the sim-to-real transferability of these predictive agents on a physical one-degree-of-freedom (1-DoF) helicopter testbed to assess their performance with regards to reference tracking.

2 Problem Formulation and Experimental Setup

The experimental setup utilizes the Quanser Aero 2 testbed in a 1-DoF configuration. We access the simulation and the physical system as described in [5]. Building upon our previous problem formulation [4], we redefine the tracking problem to better align with the dynamic constraints of physical hardware. Instead of step functions as reference signals, which are physically impossible to track for mechanical systems, we shift the target generation to continuous S-curve trajectories and constant sections. These S-curves provide physically feasible, differentiable trajectories. This prevents the agent from overfitting its policy to the unrealistic, instantaneous error jumps caused by rigid step functions, forcing it instead to learn generalized, dynamic tracking behavior. In a purely reactive setup, the environment state at time t is defined as $s_t = (\theta_t, \omega_t, r_t)$, where θ_t is the pitch angle, ω_t is the angular velocity (computed as $\Delta\theta/\Delta t$), and r_t is the current reference. To embed predictive "foresight", we augment this state space by including the target velocity $\dot{r}_t$ (computed analytically) and a horizon of future target points sampled directly from the trajectory. The augmented state space is formulated as $s_t = (\theta_t, \omega_t, r_t, \dot{r}_t, r_{t+1\cdot\Delta t}, ..., r_{t+N\cdot\Delta t})$. Here, N represents the number of future targets provided to the agent, and Δt defines the fixed time interval between these future trajectory points. To solve this control task, we selected the proximal policy optimization (PPO) algorithm due to its theoretical policy improvement guarantees and proven stability in continuous action spaces. The algorithm was implemented using the Stable Baselines3 library [3]. To isolate the impact of predictive features, we systematically evaluated eight state-space configurations formed by augmenting the purely reactive baseline with various combinations of the target velocity $\dot{r}$ and look-ahead horizons (N future targets at varying intervals Δt). (2) **Target Velocity:** augmented solely with $\dot{r}_t$; **Future Targets:** augmented with N future targets at varying intervals of Δt; and (4) **Combined:** integrating both $\dot{r}_t$ and future targets.

To ensure statistical significance, each configuration was trained in simulation over 10 independent runs using varying random seeds. The primary evaluation metric was the mean absolute deviation to the target trajectory. Finally, to evaluate the sim-to-real gap, the single best-performing agent from each configuration was deployed directly onto the physical hardware and evaluated on a single representative S-curve trajectory.

3 Results and Discussion

The experimental results of all eight evaluated configurations are summarized in Table 1. In the simulation environment, the introduction of predictive state variables yielded a roughly 9-fold improvement in tracking accuracy. The purely reactive baseline agent struggled to track the continuous S-curves, resulting in a mean absolute deviation of $2.73° \pm 0.18°$. In contrast, the fully combined predictive agent (utilizing both the target velocity $\dot{r}_t$ and $N = 4$ future targets at $\Delta t = 0.5$ s intervals) achieved near-perfect tracking with a deviation of just $0.31° \pm 0.10°$. A critical finding from the simulation data is the impact of the look-ahead horizon length. When providing the agent with only a single future target ($N = 1$), a time interval of $\Delta t = 1.0$ s significantly outperformed a short $\Delta t = 0.1$ s interval ($0.85° \pm 0.06°$ versus $1.70° \pm 0.13°$, respectively). This behavior can be attributed to the physical inertia of the 1-DoF system. A 0.1 s look-ahead does not provide sufficient time for the agent to accelerate or decelerate the beam. Conversely, a 1.0 s horizon allows the agent to smoothly "pre-steer" and apply control efforts proactively, minimizing lag. To visualize this improvement Fig. 1 illustrates the tracking behavior of the baseline agent compared to the predictive agent against the reference signal.

While the simulation results demonstrate the significant improvement of the predictive formulation, the real-world validation on the physical hardware reveals a classic sim-to-real gap. The most complex agent ($\dot{r}_t$, $N = 4$ with $\Delta t = 0.5$ s) performed best in simulation. However, during physical hardware validation, this most augmented configuration did not outperform simpler models; instead, it yielded the same top-tier performance as an agent with no target velocity and only a single future target with $\Delta t = 1.0\,s$. This phenomenon provides valuable insights into mitigating the sim-to-real gap. By providing the agent with highly granular, dense trajectory data, the neural network is able to exploit the idealized, noise-free dynamics of the simulation model. However, physical systems suffer from unmodeled friction, communication delays, and sensor noise. As observed in recent literature regarding the sim-to-real gap [6], policies that perfectly optimize for simulated dynamics often become fragile in reality. Because the simpler predictive formulation matches the hardware performance of the most complex agent, it becomes clear that highly dense trajectory data is not strictly required for optimal physical transfer. Instead, a simpler, broader look-ahead horizon acts as an effective regularizer. It provides enough foresight to overcome the system's inertia while remaining abstract enough to ignore minor dynamic discrepancies between the simulation and the real-world testbed, offering a more computationally efficient state space without sacrificing real-world accuracy.

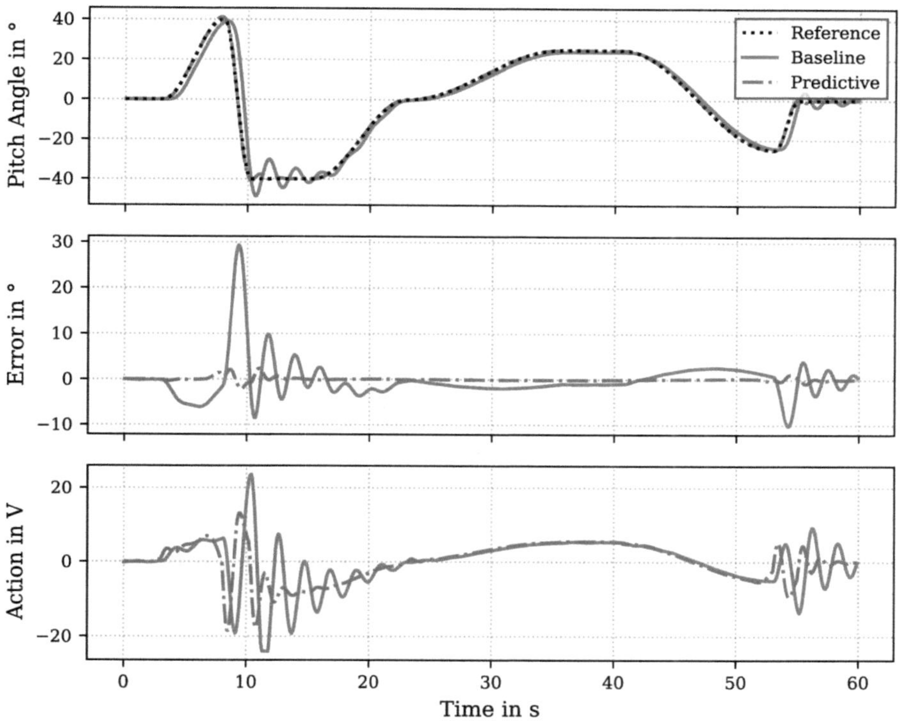

Fig. 1. Simulation evaluation profile of the reactive baseline versus the best predictive agent, showing the S-curve pitch tracking (top), target deviation error (middle), and applied control voltage (bottom).

Table 1. Comparison of tracking performance across different predictive state representations. $\dot{r}$: Target velocity included. N: Number of future target points. Δt: Time interval between future targets in seconds. Values represent the absolute pitch deviation in degrees (°). Worst, Best, Mean, and Std refer to the maximum, minimum, average, and standard deviation of the tracking error across the 10 independent simulation runs.

$\dot{r}$	N	Δt	Worst	Best	Mean	Std	Hardware
No	0	–	3.05	2.51	2.73	0.18	2.34
No	1	0.1	1.88	1.49	1.70	0.13	2.09
No	1	1.0	0.95	0.78	0.85	0.06	**1.11**
No	4	0.5	1.02	0.49	0.70	0.14	1.21
Yes	0	–	0.90	0.41	0.52	0.15	1.23
Yes	1	0.1	0.74	0.34	0.48	0.10	1.43
Yes	1	1.0	0.50	0.22	0.34	0.08	1.14
Yes	4	0.5	0.53	0.19	**0.31**	0.10	**1.11**

4 Conclusion and Future Work

We demonstrated that the performance limitations of standard DRL in physical tracking stem largely from their purely reactive problem formulation. Embedding a predictive look-ahead horizon into the state space bridges the gap between reactive AI and optimal control. This augmented formulation allows the agent to anticipate trajectory changes and compensate for physical inertia, resulting in highly accurate continuous tracking. For industry, this approach offers a highly desirable trade-off: lightweight online inference of neural networks while maintaining the high precision required to stabilize complex plants. Furthermore, sim-to-real validation highlighted that a coarser prediction horizon acts as a robust regularizer, preventing overfitting to simulated dynamics. Future work will expand this approach. First, to mitigate the reliance on *a priori* target trajectories, we will explore learning trajectory prediction models for task-specific movement patterns via supervised learning. Second, we plan fine-tuning directly on hardware and conducting statistically rigorous hardware evaluations across a broader $(N, \Delta t)$ grid. Third, we aim to scale this predictive formulation to multi-DoF robotic systems for complex manufacturing tasks. Finally, we will benchmark the real-time performance and computational overhead of our agents against advanced MPC.

Acknowledgments. Financial support for this study was provided by the Christian Doppler Research Association (CDG) through the Josef Ressel Centre for Intelligent and Secure Industrial Automation, the corresponding WISS Co-project of Land Salzburg, and by the European Interreg project BA0100172 AI4GREEN. During the preparation of this work, the authors used AI tools for language editing and formatting assistance.

References

1. Brockman, G., et al.: OpenAI gym, arXiv preprint arXiv:1606.01540 (2016)
2. Nian, R., Liu, J., Huang, B.: A review on reinforcement learning: introduction and applications in industrial process control. Comput. Chem. Eng. **139**, 106886 (2020)
3. Raffin, A., Hill, A., Gleave, A., Kanervisto, A., Ernestus, M., Dormann, N.: Stable-baselines3: reliable reinforcement learning implementations. J. Mach. Learn. Res. **22**(268), 1–8 (2021)
4. Schäfer, G., Krau, T., Rehrl, J., Huber, S., Hirlaender, S.: The crucial role of problem formulation in real-world reinforcement learning. In: 2025 IEEE 8th International Conference on Industrial Cyber-Physical Systems (ICPS), pp. 1–6. IEEE (2025)
5. Schäfer, G., Schirl, M., Rehrl, J., Huber, S., Hirlaender, S.: Python-based reinforcement learning on simulink models. In: International Conference on Soft Methods in Probability and Statistics, pp. 449–456. Springer (2024)
6. Wagenmaker, A., Huang, K., Ke, L., Jamieson, K., Gupta, A.: Overcoming the sim-to-real gap: leveraging simulation to learn to explore for real-world RL. Adv. Neural. Inf. Process. Syst. **37**, 78715–78765 (2024)

Integrating Physics-Informed Neural Networks for Safe Reinforcement Learning in a 1-DoF Helicopter System

Georg Schäfer[1,2(✉)], Jakob Rehrl[1], and Stefan Huber[1]

[1] Josef Ressel Centre for Intelligent and Secure Industrial Automation, Salzburg University of Applied Sciences, Salzburg, Austria
[2] Paris Lodron University of Salzburg, Salzburg, Austria
georg.schaefer@fh-salzburg.ac.at

Abstract. Deep reinforcement learning (DRL) offers powerful control for industrial cyber-physical systems (ICPSs), but its "black-box" exploration risks violating strict hardware safety limits. Typically, these constraints are managed through complex reward shaping. In this work-in-progress paper, we embed a differentiable physics model directly into the proximal policy optimization (PPO) actor loss function. By simulating short-horizon future trajectories during training, the policy is penalized for anticipated safety violations independent of the task-reward signal. Evaluated on a simulated 1-degree-of-freedom helicopter testbed with strict pitch constraints, our physics-informed soft regularizations substantially reduce constraint violations while maintaining reliable target tracking.

Keywords: Reinforcement Learning · Physics-Informed Neural Networks · Safe Control · Cyber-Physical Systems

1 Introduction

Industrial cyber-physical systems (ICPSs) integrate physical processes with computational control, underpinning modern manufacturing and smart production. Reinforcement learning (RL) has demonstrated significant success in optimizing these complex environments, offering data-driven solutions where mathematical modeling is challenging [4]. However, a primary limitation of standard RL is its "black-box" exploration strategy, which can cause irreversible hardware damage in real-world mechatronic systems. The target system is the Quanser Aero 2[1], a mechatronic laboratory testbed suitable for controlling the angle θ of a bar by actuating two rotors, see [7] for details. Traditionally, safety constraints in RL are encoded purely within the reward definition, leading to complex and brittle reward shaping [3]. Other safe RL paradigms rely on constrained Markov decision processes (CMDPs) [1], control barrier functions (CBFs) [2], or offline model

[1] https://www.quanser.com/products/aero-2/.

© The Author(s), under exclusive license to Springer Nature Switzerland AG 2027
M. Geiß et al. (Eds.): DEXA 2026 Workshops, CCIS 3078, pp. 107–111, 2027.
https://doi.org/10.1007/978-3-032-34202-7_10

predictive control (MPC) projections [9]. To achieve anticipatory and safe control without massive computational overhead, we propose embedding a differentiable physics model directly into the loss function of the proximal policy optimization (PPO) algorithm. This approach conditions the actor network regarding safety constraints by penalizing predicted future violations via differentiable simulation [6]. The scientific value of this work-in-progress lies in demonstrating constraint satisfaction decoupled from the primary task-reward signal, providing a scalable mechanism for safe RL deployment.

2 Methodology

Our base algorithm is PPO [11], chosen for its stable performance and continuous action space capabilities. To embed physical foresight, the system dynamics are modeled via known differential equations representing the Quanser Aero 2. We integrate the physics-informed neural network (PINN) via an auto-regressive unroll mechanism. An overview of this dual-path architecture (evaluating actions concurrently in the environment for task reward and in the differentiable physics model for safety constraints) is illustrated in Fig. 1.

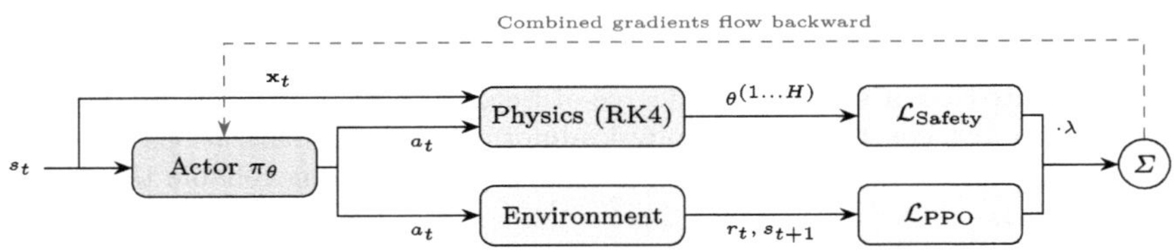

Fig. 1. Architecture overview: Actions are evaluated concurrently in the environment and the differentiable physics model. Both the task objective ($\mathcal{L}_{\text{PPO}}$) and the anticipatory safety penalty ($\mathcal{L}_{\text{Safety}}$, scaled by λ) combine into a total loss, whose gradients flow backward to update the policy.

- **Lookahead Horizon:** We project 0.3 s into the future (3 steps at 10 Hz). This balances the need to capture system inertia against the computational cost of long-term simulation.
- **System Dynamics vs. RL Observation:** For every timestep t, we distinguish the agent's full observation state s_t from the physical state $\mathbf{x}_t$ used for safety prediction. While the PPO agent observes an augmented state [8], the differentiable physics model isolates strictly the one-degree-offreedom (1-DoF) Quanser Aero 2 system state: $\mathbf{x} = [\theta, \dot{\theta}, \omega_0, \omega_1]^T$ (pitch angle, pitch velocity, and internal motor angular velocities). We use these core variables to initialize the non-linear continuous-time dynamics $\dot{\mathbf{x}} = f(\mathbf{x}, \mathbf{u})$ natively in PyTorch. During training, $\mathbf{x}_t$ is extracted directly from the simulator as privileged information, and the current policy is sampled for new actions using the predicted state as an input.

- **Numerical Integration:** We utilize a Runge-Kutta (RK4) integration method, operating at 5 micro-steps per environment step, ensuring accurate trajectory simulation.

Safety Constraint Definition: To rigorously test the mechanism, we impose an artificial pitch limit of $\pm 30°$. We simultaneously train the agent to track target signals of $\pm 40°$. While contradictory to standard operation, this extreme configuration effectively showcases the algorithm's ability to prioritize safety bounds over task-reward maximization.

Loss Modification: The standard PPO actor loss is augmented with a safety penalty derived from the differentiable physics model as $\mathcal{L} = \mathcal{L}_{\text{PPO}} + \lambda \cdot \mathcal{L}_{\text{Safety}}$, where λ acts as a weighting factor. Because the numerical RK4 integration is built entirely from trackable tensor operations, the computational graph records every simulation step. Consequently, via the chain rule, gradients from anticipated safety violations flow backward through the unrolled differential equations into the actor network. To compute $\mathcal{L}_{\text{Safety}}$, we unroll the system dynamics over the prediction horizon of $H = 3$ steps. Let $\theta^{(h)}$ denote the predicted pitch angle at horizon step h, simulated via the differentiable RK4 integrator. The safety penalty acts as a soft regularizer, relying on a rectified linear unit (ReLU) to penalize the policy strictly when the absolute predicted pitch exceeds the predefined safety limit $\theta_{\max} = 30°$. The normalized safety loss is defined as:

$$\mathcal{L}_{\text{Safety}} = \frac{1}{H} \sum_{h=1}^{H} \max(0, \frac{|\theta^{(h)}|}{\theta_{\max}} - 1) \tag{1}$$

Because this formulation is purely composed of differentiable tensor operations, the resulting gradients correctly map anticipated future constraint violations back to the actor network's current action distribution.

3 Experimental Setup and Results

We evaluate our approach on a simulated Quanser Aero 2 system in a 1-DoF configuration. The underlying system dynamics and base RL problem formulation have been established in previous studies [7], and the environment is accessed via a standard Gym API framework using Python-Simulink interfaces [10]. We tested three primary configurations, running 10 random seeds for each to ensure statistical significance: *Naive Baseline* ($\lambda = 0$), *PINN Balanced* ($\lambda = 150$), and *PINN Over-penalized* ($\lambda = 1000$). The penalty weight λ was chosen empirically; a rigorous statistical evaluation of all 10 seeds and a λ Pareto sweep are reserved for future work.

Training Stability: All configurations successfully converged. We observed that the value loss naturally increases for higher λ configurations due to the conflicting objectives (tracking a 40° target vs. stopping at 30°), but the policy loss remained comparable across the board.

Performance Trade-Off and Safety: The evaluation results (see Fig. 2) highlight a clear trade-off between target tracking and constraint satisfaction. The *Naive Baseline* achieves high task reward by accurately tracking the 40° target, but heavily violates the 30° safety threshold. The *PINN Over-penalized* model exhibits very few safety violations but results in sluggish overall tracking, failing to adapt dynamically even when within safe regions. The *PINN Balanced* configuration demonstrates a suitable choice of λ. It aggressively tracks the reference signal and largely learns to decelerate to respect the 30° threshold. Although brief transient violations can still occur during abrupt, extreme setpoint changes, it drastically reduces both predicted and actual pitch violations compared to the baseline.

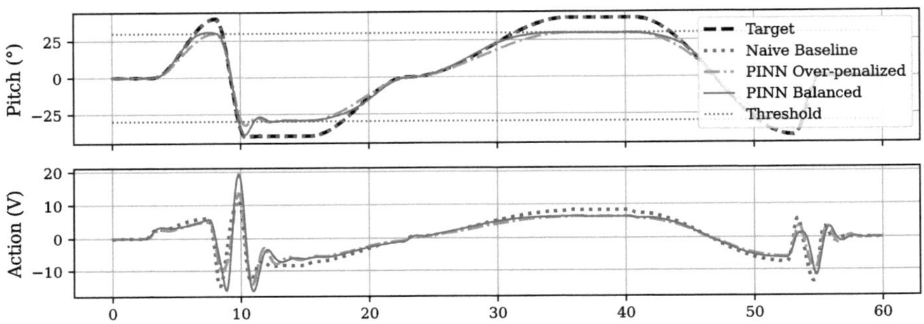

Fig. 2. Simulation evaluation profile demonstrating the trade-off between target tracking and constraint satisfaction. The *Naive Baseline* ignores the 30° safety threshold, the *PINN Over-penalized* struggles to track the signal, and the *PINN Balanced* successfully adheres to the constraints while maintaining strong tracking performance.

4 Conclusion and Future Work

In this work-in-progress paper, we demonstrated that embedding differentiable physics models into the PPO actor loss function provides a scalable and effective mechanism for soft anticipatory safety regularization in ICPS. This approach enables anticipatory constraint satisfaction and mitigates the need for complex, task-specific reward shaping. Ongoing and future research will focus on the sim-to-real transfer of this PINN-trained agent directly onto the physical Quanser Aero 2 hardware. Beyond zero-shot transfer, we plan to conduct further fine-tuning directly on the real hardware to close the sim-to-real gap. Because relying on a static ordinary differential equation (ODE) model during this phase could introduce systematic errors, this fine-tuning should be coupled with active parameter identification. Building upon our recent work on parameter identification methodologies for the testbed [5], we plan to dynamically update the parameters of the differentiable physics model to match observed real-world trajectories. Finally, we aim to complement our anticipatory training loss with a reactive safety mechanism (such as a safety shield or CBF) that acts as a real-time execution filter. While the predictive loss smoothly guides the policy toward safe behaviors during training, a reactive architectural constraint would intervene directly during deployment, instantly overriding proposed actions if a forward-simulation predicts an imminent pitch violation.

Acknowledgments. Financial support for this study was provided by the Christian Doppler Research Association (CDG) through the Josef Ressel Centre for Intelligent and Secure Industrial Automation, the corresponding WISS Co-project of Land Salzburg, and by the European Interreg project BA0100172 AI4GREEN. During the preparation of this work, the authors used AI tools for language editing and formatting assistance.

References

1. Altman, E.: Constrained Markov Decision Processes. Routledge (2021)
2. Ames, A.D., Coogan, S., Egerstedt, M., Notomista, G., Sreenath, K., Tabuada, P.: Control barrier functions: theory and applications. In: 18th European Control Conference (ECC), pp. 3420–3431. IEEE (2019)
3. Garcıa, J., Fernández, F.: A comprehensive survey on safe reinforcement learning. J. Mach. Learn. Res. **16**(1), 1437–1480 (2015)
4. Kober, J., Bagnell, J.A., Peters, J.: Reinforcement learning in robotics: a survey. Int. J. Robot. Res. **32**(11), 1238–1274 (2013)
5. Langschwert, J., Schäfer, G., Rehrl, J., Huber, S., Hirlaender, S.: Reinforcement learning for optimal experiment design in parameter identification of mechatronic systems. In: Database and Expert Systems Applications - DEXA 2026 Workshops. CCIS. Springer, Cham, Graz, Austria (2026, submitted)
6. Raissi, M., Perdikaris, P., Karniadakis, G.E.: Physics-informed neural networks: a deep learning framework for solving forward and inverse problems involving nonlinear partial differential equations. J. Comput. Phys. **378**, 686–707 (2019)
7. Schäfer, G., Rehrl, J., Huber, S., Hirlaender, S.: Comparison of model predictive control and proximal policy optimization for a 1-DOF helicopter system. In: 2024 IEEE 22nd International Conference on Industrial Informatics (INDIN'24). IEEE, Beijing, China (2024). https://doi.org/10.1109/INDIN58382.2024.10774357
8. Schäfer, G., Rehrl, J., Huber, S., Hirlaender, S.: Anticipatory reinforcement learning for trajectory tracking. In: Database and Expert Systems Applications - DEXA 2026 Workshops. CCIS. Springer, Cham, Graz, Austria (2026, submitted)
9. Schäfer, G., Rehrl, J., Huber, S., Hirlaender, S.: Safe reinforcement learning using ideas from model predictive control. In: Computer Aided Systems Theory – EUROCAST 2026 – Extended Abstracts, vol. 20 (2026)
10. Schäfer, G., Schirl, M., Rehrl, J., Huber, S., Hirlaender, S.: Python-based reinforcement learning on simulink models. In: 11th International Conference on Soft Methods in Probability and Statistics (SMPS 2024). Salzburg, Austria (2024). https://doi.org/10.1007/978-3-031-65993-5_55
11. Schulman, J., Wolski, F., Dhariwal, P., Radford, A., Klimov, O.: Proximal policy optimization algorithms. arXiv preprint arXiv:1707.06347 (2017)

Reinforcement Learning for Optimal Experiment Design in Parameter Identification of Mechatronic Systems

Julian Langschwert[1,2(✉)], Georg Schäfer[1,2], Jakob Rehrl[1], Stefan Huber[1], and Simon Hirländer[2]

[1] Josef Ressel Centre for Intelligent and Secure Industrial Automation, University of Applied Sciences, Salzburg, Austria

[2] Paris Lodron University of Salzburg, Salzburg, Austria

julian.langschwert@fh-salzburg.ac.at

Abstract. Informative excitation signals are critical for accurate system identification of mechatronic systems, yet classical system identification (SI) approaches require expert knowledge and hand-crafted signal design to respect hardware safety constraints, limiting their generalizability. We propose a RL agent that learns optimal excitation signals for a Quanser Aero 2 testbed while autonomously enforcing safety constraints through reward shaping. Evaluated across 10 independent training seeds, our comprehensive agent achieves competitive estimation accuracy across all three identified parameters, outperforming classical baselines while incurring only 0.75% safety violations.

Keywords: Reinforcement Learning · System Identification · Optimal Experimental Design · Cyber-Physical Systems

1 Introduction

SI is a fundamental process for modeling complex dynamics [3]. Field experts with extensive system and domain knowledge are required for tasks such as model selection, optimal experimental design (OED), data collection, and parameter estimation. A complicating factor is that many cyber-physical systems (CPSs) have strict hardware or safety constraints that need to be respected when designing these experiments. Current SI strategies rely on standard open-loop excitation signals such as pseudo-random binary sequences (PRBS) or frequency sweeps [3]. While highly effective, their application necessitates extensive manual tuning to simultaneously maximize information yield [4] and respect strict physical safety constraints. Recent works demonstrate that reinforcement learning (RL) can successfully automate OED; for instance, by optimizing input sequences for parameter estimation in Lithium-ion batteries [2], and by utilizing D-optimality-guided rewards to efficiently calibrate a multi-DOF rehabilitation robot [1]. Building upon this concept, this work-in-progress paper contributes an

© The Author(s), under exclusive license to Springer Nature Switzerland AG 2027
M. Geiß et al. (Eds.): DEXA 2026 Workshops, CCIS 3078, pp. 112–117, 2027.
https://doi.org/10.1007/978-3-032-34202-7_11

RL agent that learns an excitation signal for a Quanser Aero 2 testbed in a 1-degree-of-freedom (DOF) configuration while minimizing parameter estimation error and autonomously respecting hardware safety constraints through reward shaping.

2 Methodology and Experimental Setup

System Description and Model Formulation. The Quanser Aero 2 is a dual-rotor testbed configured here for a 1-DOF pitch motion (yaw locked, rotors horizontal). It is driven by a single voltage input $u \in [-24, 24]$ V, applied to the motors as $u_1 = u$ and $u_2 = -u$, resulting in a seesaw motion with a manufacturer-specified pitch limit of $\pm 45°$.

The dynamic behavior of the pitch axis is governed by a second-order nonlinear differential equation:

$$J_p\ddot{\Theta} = (F_1 - F_2)l - D_p\dot{\Theta} - mgd_S \sin(\Theta) \quad (1a)$$

$$F_1 = \frac{J_p}{l}K_{pu}u_1, \quad F_2 = \frac{J_p}{l}K_{pu}u_2 \quad (1b)$$

where Θ, $\dot{\Theta}$, and $\ddot{\Theta}$ represent the pitch angle, angular velocity, and angular acceleration, respectively. The terms F_1 and F_2 denote the thrust forces generated by the rotors, which are located at a distance l from the pivot. While physical constants such as the system mass (m), gravity (g), and center of mass offset (d_S) are known, three core dynamic parameters remain unknown and are the target of our SI: the moment of inertia (J_p), the viscous damping coefficient (D_p), and the voltage-to-thrust gain (K_{pu}).

Optimal Experiment Design via Reinforcement Learning. We formulate the active excitation design as a finite-horizon Markov decision process (MDP) defined by the tuple $(\mathcal{S}, \mathcal{A}, r, \gamma)$.

State and Action Space: The agent observes a sliding window of the $N_w = 80$ most recent noisy angle measurements and applied voltages. It additionally receives a normalized time index k/T to track progression across the $T = 500$ step episode, enabling the agent to adapt its excitation strategy over time:

$$s_k = \left[\frac{y_{k-N_w+1}}{y_{\max}}, \ldots, \frac{y_k}{y_{\max}}, \frac{u_{k-N_w+1}}{u_{\max}}, \ldots, \frac{u_k}{u_{\max}}, \frac{k}{T}\right] \quad (2)$$

At each step k, the agent outputs a continuous action $a_k \in [-1, 1]$, scaled to the physical motor voltage limits of ± 24 V.

Parameter Estimation: For parameter identification, a discrete AutoRegressive with eXogenous inputs (ARX) model coupled with a (RLS) estimator is utilized to recover the physical parameters via a nonlinear inverse mapping at the end of the episode.

Reward Formulation: The objective of the agent is to maximize the expected discounted return. The reward function is designed to balance parameter estimation quality with strict hardware safety, quantified by:

1. Stepwise Safety Penalty: The Quanser Aero 2 has a manufacturer-specified operating range of $\pm 45°$. To account for system inertia and avoid mechanical stress at the physical boundaries, we define a conservative critical limit $\theta_{\text{lim}} = 40°$. Furthermore, a warning threshold of $\theta_{\text{warn}} = 30°$ was determined empirically to provide a sufficient $10°$ buffer zone, allowing the agent to learn to decelerate before breaching the limit. The agent receives a quadratic penalty inside this warning zone and a severe penalty for breaching the critical limit:

$$r_k^{\text{safety}} = \begin{cases} -1 - 2\frac{|\theta_k| - \theta_{\text{lim}}}{\theta_{\text{lim}}} & \text{if } |\theta_k| \geq \theta_{\text{lim}}, \\ -0.1\left(\frac{|\theta_k| - \theta_{\text{warn}}}{\theta_{\text{lim}} - \theta_{\text{warn}}}\right)^2 & \text{if } |\theta_k| \geq \theta_{\text{warn}}, \\ 0 & \text{otherwise.} \end{cases} \tag{3}$$

2. Terminal Estimation Reward: Crucially, the (RL) agent is trained entirely in a simulation environment. During this training phase, the physical parameters are based on nominal estimates from prior baseline experiments. In simulation, these parameters serve as the known ground truth p_i^* for the reward calculation, allowing the agent to learn an optimal policy offline that can subsequently be deployed on real hardware where the parameters are truly unknown. Because the estimation quality is evaluated at the end of the trajectory, the agent receives a sparse terminal reward based on the relative error between the (RLS)-estimated parameters $\hat{p}_{i,T}$ and the simulated ground truth p_i^*:

$$r_T^{\text{est}} = - \sum_{p_i \in \mathcal{P}_{\text{target}}} \frac{|\hat{p}_{i,T} - p_i^*|}{p_i^*} \tag{4}$$

Training and Evaluation Setup. The (RL) agent uses a proximal policy optimization (PPO) algorithm from Stable-Baselines3 with a default net architecture of two hidden layers of 64 neurons each. To ensure robust learning and address the sim-to-real gap, the true physical parameters are randomized by $\pm 30\%$ of their nominal values at the start of each episode. To thoroughly evaluate the proposed approach, four different (RL) agents were trained. Three specialized agents were trained to prioritize the estimation accuracy of a single specific parameter (J_p, K_{pu}, or D_p). While the underlying (RLS) algorithm always estimates all three parameters, this modification was achieved by changing the terminal reward to penalize only the relative error of the targeted parameter. A fourth, comprehensive agent was trained to jointly estimate all three parameters using the full terminal reward formulation. Each agent was trained with ten independent random seeds to ensure robustness to training stochasticity.

Baseline Comparison: To evaluate the (RL) approach, the trained agents are benchmarked against three classical open-loop signals, all restricted to a conservative $\pm 7.2\,\text{V}$ amplitude to ensure hardware safety. The first is a naive **random** excitation drawn uniformly from $[-7.2\,\text{V}, 7.2\,\text{V}]$ at each step. The second is a persistently exciting **PRBS** switching between discrete $\pm 7.2\,\text{V}$ states. The third is a **Chirp** signal sweeping from 0.05 to 2 Hz, providing broadband excitation tailored to the pitch dynamics. For a strictly fair comparison, all methods are evaluated at a 10 Hz sampling rate over the exact same episode length using the identical least-squares algorithm.

3 Results and Discussion

Training Stability. As detailed in Table 1, the N_{seeds} column indicates that targeting D_p makes training less stable. While the parameter-specific agents RL(J_p) and RL(K_{pu}) achieved full convergence, RL(D_p) converged in 9/10 seeds, and RL(all) in only 8/10. Specifically, two of the comprehensive agents collapsed to a zero-output policy, highlighting the challenge of jointly optimizing for all three parameters simultaneously. Furthermore, the observed standard deviations (SDs) are generally large. For the J_p estimation of the RL(K_{pu}) agent, the SD entirely exceeds the mean, underlining the need for more stable training procedures in future work.

Estimation Accuracy and Safety Trade-Off. The RL(D_p) agent outperforms all other policies across *all three* parameters, suggesting that successfully exciting the most difficult parameter yields a trajectory rich enough to identify the remaining ones. However, the RL(D_p) agent's regular safety violations of 151/900 evaluations (a rate of approximately 16.8%) render it unsuitable for deployment on physical hardware. Consequently, the comprehensive agent emerges as the practically preferred policy, achieving D_p error of 198% versus 495% for the best classical baseline while safely operating within physical limits, incurring only 0.75% safety violations.

Damping Identification Challenges. Despite these results, all methods continue to struggle with accurately identifying the viscous damping coefficient D_p. This is consistent with ill-conditioning in the discrete AutoRegressive with eXogenous inputs (ARX) regressor matrix: D_p contributes weakly to the regression signal relative to J_p and K_{pu}, making its contribution difficult to isolate regardless of excitation quality.

Actuator Range and Comparison Validity. A key advantage of the RL approach is its ability to dynamically utilize the full $\pm 24\,\text{V}$ range while respecting safety constraints through reward shaping. Classical baselines required manual restrictions to $\pm 7.2\,\text{V}$ to prevent safety violations. Yet, even with this conservative amplitude, the PRBS baseline still exceeded the physical limit twice. While this disparity in operating voltage limits direct comparison, it simultaneously demonstrates the core contribution of RL-based OED: rather than requiring conservative hand-crafted amplitude limits, the agent learns to safely exploit the available actuator range to maximize information yield.

Conclusion and Future Work. This work demonstrates that RL can successfully automate optimal experimental design for mechatronic systems by learning excitation signals that outperform classical baselines. However, accurately identifying D_p remains an open challenge. To address this, future work will replace

Table 1. Robustness across 10 training seeds (N_{seeds} denotes viable seeds), evaluated on 100 randomized episodes. Values report the cross-seed mean ± std of the per-seed mean absolute relative error (MARE) of parameters and mean absolute maximum angles during evaluation.

Policy	J_p / %	K_{pu} / %	D_p / %	Safety Viol.	$\theta_{\max}$ / °	N_{seeds}
Baselines *(±7.2 V)*						
Random	8.23 ± 6.61	10.18 ± 5.14	3026 ± 2067	0/100	11.2 ± 4.6	—
PRBS	2.57 ± 2.73	3.42 ± 1.64	1065 ± 936	2/100	20.4 ± 9.1	—
Chirp	1.43 ± 0.57	2.65 ± 1.09	495 ± 202	0/100	24.4 ± 4.3	—
RL Agents *(±24 V, cross-seed statistics)*						
RL(J_p)	1.87 ± 1.38	1.86 ± 1.15	4577 ± 2436	6/1000	17.0 ± 4.7	10/10
RL(K_{pu})	5.70 ± 5.79	0.40 ± 0.36	1727 ± 1179	3/1000	16.9 ± 5.6	10/10
RL(D_p)	**0.75 ± 0.38**	**0.37 ± 0.20**	**190 ± 86**	151/900	36.1 ± 6.0	9/10
RL(all)	1.54 ± 1.04	0.51 ± 0.40	198 ± 136	6/800	26.5 ± 2.0	8/10

the discrete ARX-RLS estimator with a batch Nonlinear Least Squares (NLS) approach operating on the continuous-time trajectory, avoiding ill-conditioning and replace the sparse terminal reward with a dense, step-wise reward formulated around the Fisher Information Matrix, which directly measures the information content of the excitation trajectory. Furthermore the safety formulation will adapt Lagrangian constraints to dynamically balance excitation and safety penalties during training. After investigating broader parameter randomization bounds to improve policy robustness, and conducting a formal optimization of the classical baseline parameters, the learned policies will be validated on the physical Quanser Aero 2 hardware. A control barrier function (CBF) will be tested to guarantee absolute hardware safety during sim-to-real transfer.

Acknowledgments. Financial support for this study was provided by the Christian Doppler Research Association (CDG) through the Josef Ressel Centre for Intelligent and Secure Industrial Automation, the corresponding WISS Co-project of Land Salzburg, and by the European Interreg project BA0100172 AI4GREEN. Google Gemini 3.1 Pro was utilized during this work to assist with software development, control architecture design, and copy-editing select paragraphs.

References

1. Hu, Q., Celler, B., Mu, W., Su, S.W.: D-Optimality-Guided Reinforcement Learning for Efficient Open-Loop Calibration of a 3-DOF Ankle Rehabilitation Robot (Jan 2026). https://doi.org/10.48550/arXiv.2601.15707
2. Huang, R., Fogelquist, J., Lin, X.: Reinforcement learning of optimal input excitation for parameter estimation with application to li-ion battery. IEEE Trans. Industr. Inf. **19**(11), 11160–11170 (2023). https://doi.org/10.1109/TII.2023.3244342

3. Ljung, L.: System identification (2nd ed.): theory for the user. Prentice Hall PTR, USA (1999)
4. Mazhar, M.F., Wasim, M., Abbas, M., Riaz, J., Swati, R.F.: Aircraft system identification using multi-stage PRBS optimal inputs and maximum likelihood estimator. Aerospace **12**(2), 74 (2025). https://doi.org/10.3390/aerospace12020074

Reconstructing OPC UA Address Spaces from Time-Series Databases

Lukas Lürzer(✉), Hannes Unger, and Stefan Huber

Josef Ressel Centre for Intelligent and Secure Industrial Automation, Salzburg University of Applied Sciences, Salzburg, Austria
{lukas.luerzer,hannes.unger,stefan.huber}@fh-salzburg.ac.at

Abstract. OPC UA has become the dominant open protocol in operational technology. Time-series databases routinely archive OPC UA telemetry but discard the semantic metadata (node hierarchy, engineering units, and type definitions) which gives sensor values their meaning. Recovering this information from a time-series database is non-trivial: namespace indices recorded at the source are session-local and unstable across restarts, and naive merging across multiple source servers results in identifier collisions. We present *opcua-ts*, an implemented architecture that persists this semantic information alongside its telemetry in a general-purpose time-series database under a lifecycle-stable join key, and that reconstructs the source address space as a live OPC UA endpoint. We characterize the conditions under which the reconstruction is sound across multi-source deployments and validate the approach with a NodeSet2 XML round-trip against the source server. Initial results from a boiler-simulator round-trip indicate that the approach is feasible.

Keywords: OPC UA · Information Model · Time-Series Databases · Semantic Metadata · Address Space Reconstruction

1 Introduction

OPC UA [3] has become the dominant protocol for machine-to-machine communication in operational technology (OT). Every compliant server exposes an *address space*: a typed, hierarchical graph of objects and variables with engineering units, data types, references, and browse paths. Additionally, the OPC Foundation maintains a growing family of standardized *companion specifications* (for robotics, machine tools, process control, energy, and other domains), so that compliant servers across vendors expose semantically interoperable type systems out of the box.

In practice this information model is usually lost at the OT/IT boundary, because typical ingestion pipelines (Telegraf, MQTT bridges, ad-hoc adapters) marshal each value into a schema-light row keyed by node identifier, and the surrounding structural metadata has no natural home there. Downstream analytics, dashboards, and ML pipelines therefore operate on context-free numeric

© The Author(s), under exclusive license to Springer Nature Switzerland AG 2027
M. Geiß et al. (Eds.): DEXA 2026 Workshops, CCIS 3078, pp. 118–123, 2027.
https://doi.org/10.1007/978-3-032-34202-7_12

streams and reintroduce the missing semantics through hand-maintained mapping tables, which drift out of sync as the source address space evolves and the mapping is updated only when someone notices.

Recovering the address space from such a time-series store is non-trivial. The binding between values and their semantic context is broken at ingestion, namespace indices recorded in the source server are session-local and unstable across server restarts, and naive merging across multiple source servers admits identifier collisions in both the namespace and the browse-name dimension. We therefore investigate the following research questions:

- **RQ1.** Can the OPC UA information model be persisted alongside its telemetry in a general-purpose time-series database with a join key?
- **RQ2.** Under what conditions can the original address space be reconstructed from such a store and served as a live OPC UA endpoint, including across multi-source deployments where namespace and browse-name collisions are unavoidable?
- **RQ3.** How can the semantic fidelity of a reconstructed address space be validated against the original source?

To answer these questions, we contribute (i) a join key derived from the namespace *URI* rather than the session-local namespace index, which makes it canonical (one form per node) and lifecycle-stable (unchanged across server restarts and crawl boundaries), and under which telemetry and semantic metadata can be co-located in a general-purpose time-series store; (ii) an aggregate reconstruction server that builds the persisted information model into a live OPC UA endpoint and manages namespace and browse-name collisions across multiple sources; and (iii) a validation procedure based on a NodeSet2 [3] round-trip against the source server.

2 Related Work

Historical access to OPC UA data is standardized through the Historical Access profiles defined in OPC UA Part 11 [2], and a number of commercial historians implement them. These products treat the address space as a runtime structure that is mirrored at configuration time, not as a first-class artifact persisted alongside the time series; an address space recovered from such a historian therefore reflects the configuration snapshot at mirror time rather than the lifecycle of the source server.

Semantic-aware digital-twin platforms address the closely related problem of giving industrial values a queryable semantic context, but typically do so through proprietary graph or semantic stores attached to the time-series layer via integrations, rather than by co-locating the semantic and temporal layers under a shared key. To our knowledge, no open system persists OPC UA semantic metadata in a general-purpose time-series database such that the source address space can be reconstructed programmatically.

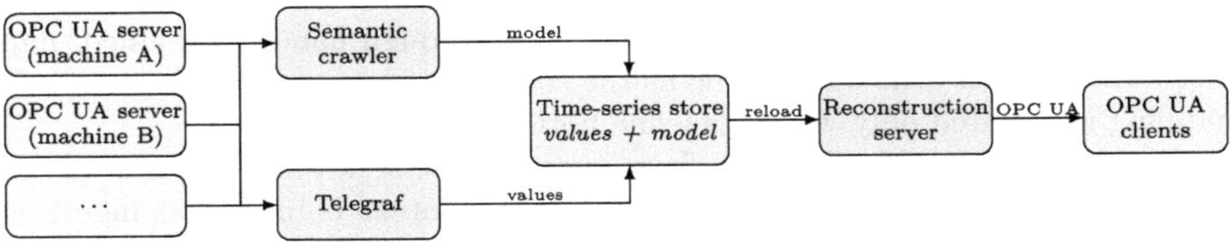

Fig. 1. Architecture proposal of *opcua-ts*: a semantic crawler and Telegraf persist the information model and live values under a shared join key; a reconstruction server serves the aggregate as a live OPC UA endpoint.

3 System Architecture

Figure 1 shows the workflow. Two ingestion paths feed the store under a shared `node_id` tag: a *semantic crawler* persists per-node metadata, while *Telegraf* [1] polls live values. The design is source-agnostic: any compliant OPC UA server (PLC-embedded, gateway, simulator) can be used, and the reconstruction server reproduces its information model without requiring access to the original device.

3.1 Storage Model

opcua-ts co-locates two complementary stores in the same time-series database, joined by a shared key: a *values store* carrying live telemetry, written by Telegraf, and a *metadata store* carrying the semantic information model, written by the crawler. The two stores share a canonical join key `node_id` (its grammar appears in Listing 1.1). The key is derived from the namespace *URI* rather than the namespace *index*: indices are session-local and unstable across server restarts, whereas URIs are durable across the lifecycle of the source server.

The metadata store carries, per node, the display name, browse path, datatype name, type-definition name, parent `node_id`, engineering unit, `EURange`, non-hierarchical references, and a monotonic `crawl_version` field that enables both `last()` snapshots and historical change detection without overwriting prior rows. Listing 1.1 shows the join-key grammar and a representative metadata record.

3.2 Crawler Pipeline

The crawler performs a depth-first traversal of the source server starting from the standard `Objects` (`i=85`) and `Types` (`i=86`) roots, recording for each node the hierarchical references used to reach it and the `HasTypeDefinition` target. A second pass enriches `Variable` nodes with `EURange` and `EngineeringUnits` via the `HasProperty` reference.

Listing 1.1. Canonical `node_id` and a sample `opcua_metadata` row (selected fields).

```
# Canonical node_id grammar
<namespace_uri>|<id_type>|<identifier>
# Example: \url{http://myplant.com/boiler/UA/|i|1002}

# Sample opcua_metadata row (selected fields):
node_id : \url{http://myplant.com/boiler/UA/|i|1002}
display_name : Loop Temperature
browse_path : /Objects/Boiler1/BoilerLoop1/LoopTemperature
data_type : Double
type_def : AnalogItemType
eu_range : -50:200
engineering_unit : degree Celsius
parent_node_id : \url{http://myplant.com/boiler/UA/|i|1001}
crawl_version : 42
```

3.3 Reconstruction Pipeline

The reconstruction server loads the latest metadata rows from the metadata store into a node-opcua [4] address space. It employs two collision-resolution mechanisms for multi-source deployments: when two source servers share a namespace URI, the aggregate registers each instance with a `#source=` fragment suffix so that namespace indices remain distinct; and a router places one container object per source endpoint under `Objects`, isolating browse-name collisions. Each `Variable` is wired to a value cache keyed on `endpoint::node_id`, populated by a background poll against the values store, so clients see the most recent value with its original `SourceTimestamp`.

4 Validation via NodeSet2 Round-Trip

We say a reconstructed address space is *semantically faithful* to its source if every browse, read, and type-resolution observation a standard OPC UA client may perform yields the same answer against the reconstruction as against the source, and if the per-namespace NodeSet2 XML serialization preserves `BrowseName`, `NodeClass`, `DataType`, `TypeDefinition`, browse-path position, engineering metadata, and non-hierarchical references node by node. We assess this by exporting both the source and the reconstructed server's address spaces to NodeSet2 XML and comparing them attribute by attribute.

NodeSet2 XML is the canonical OPC UA address-space serialization defined in the OPC UA specifications [3], and the node-opcua library provides both a loader (`generateAddressSpace`) and a per-namespace serializer (`toNodeset2XML()`). We rely on this serializer for both directions of the round-trip.

Procedure. The round-trip proceeds in four steps. First, we export the source server's namespace to NodeSet2 XML directly via `toNodeset2XML()`. Second, we run the crawler against the same source server, populating the metadata store with one row per node. Third, we instantiate a reconstruction from the latest crawl. Finally, we export the reconstruction's namespace using the same serializer and compare the two NodeSet2 files attribute by attribute.

Initial Findings. We exercised the round-trip on a boiler-simulator address space with eleven functional nodes, including five `AnalogItemType` variables. The source serialized to 25.4 KB and the reconstruction to 19.9 KB. All functional nodes survive the round-trip together with their data types, type definitions, the `EURange`, `EngineeringUnits`, and `InstrumentRange` properties of the analog items, and the non-hierarchical `GeneratesEvent` reference. This indicates that the approach is feasible in principle.

Known Gaps. The 5.5 KB size delta between the two files is fully accounted for by two optional display hints, `Definition` and `ValuePrecision`. The metadata loader filters all `PropertyType` sub-nodes from the snapshot, because node-opcua's `addAnalogDataItem()` re-emits the `EURange`, `EngineeringUnits`, and `InstrumentRange` properties from the primary metadata fields, and re-loading the source's own property nodes would collide. The two hints are dropped as collateral; widening the metadata query to preserve them is straightforward future work.

5 Discussion and Conclusion

Coupling the semantic and temporal layers in a single store lets analytics pipelines join telemetry with engineering units and browse paths without an external mapping table, allows the reconstruction server to expose a semantically-rich OPC UA endpoint to clients that cannot query the time-series store directly, and makes the appearance or disappearance of a `node_id` (via `crawl_version`) a queryable event. Limitations remain: the reconstruction is snapshot-based, and custom `ObjectType`/`VariableType` definitions reattach to built-in base types in the current implementation. OPC UA semantic metadata can therefore be persisted alongside its telemetry without sacrificing the ability to reconstruct the source address space. Future work covers type-hierarchy preservation and subscription-driven crawls.

Acknowledgments. During the preparation of this work, the authors used AI tools for language editing and formatting assistance.

References

1. InfluxData: Telegraf OPC UA Input Plugin (2024). https://github.com/influxdata/telegraf/tree/master/plugins/inputs/opcua
2. OPC Foundation: OPC UA Specification, Part 11 (Historical Access). Specification Release 1.05, OPC Foundation (2023)
3. OPC Foundation: OPC UA Specification, Parts 1 (Overview) and 6 (Mappings). Specification Release 1.05, OPC Foundation (2023)
4. Rossignol, E.: node-opcua—OPC UA stack for Node.js (2024). https://github.com/node-opcua/node-opcua

Self-supervised CAD-Face Localization in Technical Drawings

Manuel Götz[1(✉)], Max Grieshammer[2], Philipp Kellner[2], and Frank Schirmeier[1]

[1] Institute for Data-Optimized Manufacturing (IDF), University of Applied Sciences Kempten, Bahnhofstr. 61, 87435 Kempten, Germany
manuel.goetz@hs-kempten.de

[2] PartSpace GmbH, Ulrichsbergerstr. 17, 94469 Deggendorf, Germany
m.grieshammer@partspace.io

Abstract. In manufacturing, 3D CAD models define the geometry of a part while 2D technical drawings specify its manufacturing requirements, yet the two are rarely linked automatically. Associating drawing annotations with their corresponding geometric surfaces in the 3D model is a prerequisite for automated Product Manufacturing Information (PMI) extraction and remains unsolved. We formulate this as self-supervised CAD-face localization: given a point marker on a projected surface in a technical drawing, the model identifies the corresponding face in the 3D CAD model. In deployment, such markers are obtained from PMI arrow tips, making our approach a composable component of an end-to-end PMI transfer pipeline. A Graph Neural Network encodes 3D geometry while a CNN or vision transformer encodes 2D drawing crops, fused via cross-attention. Training uses synthetic data of varying complexity, requiring no manual annotation. Our best configuration achieves 98.04% accuracy on the face matching task, demonstrating successful cross-modal correspondence learning.

Keywords: CAD · Technical Drawings · Cross-Modal Learning · Graph Neural Networks · Self-Supervised Learning

1 Introduction

The manufacturing industry is undergoing a deep digital transformation, driven by the need to make product data machine-readable, interoperable, and traceable across the entire product lifecycle. At the heart of this transformation lies the challenge of bridging two coexisting but largely disconnected representations of a part: the 3D CAD model, which encodes geometry and topology, and the 2D technical drawing, which contains manufacturing intent in the form of dimensions, tolerances, surface specifications, and geometric dimensioning and tolerancing (GD&T) symbols, collectively referred to as Product Manufacturing Information (PMI). Despite decades of CAD standardization, the technical drawing remains the dominant carrier of PMI in industrial practice. These drawings are typically authored and consumed by human experts, and their machine readability remains severely limited.

© The Author(s), under exclusive license to Springer Nature Switzerland AG 2027
M. Geiß et al. (Eds.): DEXA 2026 Workshops, CCIS 3078, pp. 124–136, 2027.
https://doi.org/10.1007/978-3-032-34202-7_13

The manual association of PMI annotations from 2D drawings to their corresponding geometric entities in the 3D model is a time-consuming, error-prone process that represents a significant bottleneck in modern manufacturing workflows. Automating this process would not only reduce engineering effort and improve data consistency, but would also enable downstream applications such as automated process planning, tolerance analysis and quality control [1].

Recent advances in deep learning have opened new avenues for automating such tasks, particularly through learned representations of both geometric structures and visual documents. However, the joint understanding of 3D CAD geometry and 2D engineering drawings remains a largely unsolved problem, owing to the semantic gap between the two modalities, the complexity of engineering drawing conventions, and the scarcity of annotated real-world training data. We address the annotation scarcity directly by deriving ground-truth correspondences automatically from CAD geometry (see Sect. 3.2), requiring no human labeling. We use the term self-supervised in this broader sense, referring to methods that generate supervision programmatically rather than through manual annotation.

1.1 Related Work

GNNs for B-Rep and CAD Geometry. The boundary representation (B-Rep) is the standard data format for solid CAD models, encoding geometry through a hierarchy of faces, edges, and vertices. Several works have proposed graph-based learning directly on B-Rep structures. UV-Net [2] introduced a CNN-based face encoder operating in the parametric UV domain, enabling expressive per-face embeddings that capture fine-grained surface geometry. BRepNet [3] and subsequently Hierarchical CADNet [4] proposed graph neural network architectures over B-Rep topology for segmentation and classification tasks. GATv2 [5] introduced dynamic graph attention, making it well-suited for heterogeneous face neighborhoods in B-Rep graphs.

Technical Drawing Understanding. Automated interpretation of engineering drawings has a long history in document analysis, initially focused on rule-based symbol recognition and line extraction. Recent deep learning approaches apply object detection and transformer-based document understanding models to extract structured information like PMI directly from drawing images [6, 7]. A related line of work reconstructs 3D CAD models from 2D orthographic projections using vision-language models [8], demonstrating promising 2D-to-3D geometric understanding. However, all these methods either treat the drawing in isolation or focus on geometric reconstruction, none establishes explicit correspondences between drawing annotations and individual B-Rep faces, which is the problem we address.

Cross-Modal Matching. The problem of associating entities across two different modalities has been extensively studied in vision-language research, with frameworks such as CLIP [9] demonstrating that strong cross-modal representations can be learned from paired data. A related line of work addresses sketch-to-3D shape retrieval, where a 2D sketch or silhouette is matched against a database of 3D shapes using learned cross-modal embeddings [10, 11]. While sharing the broad goal of bridging 2D and 3D representations, these methods operate on whole-shape retrieval from freehand sketches rather than fine-grained face-level correspondence from structured engineering drawings. Transformer-based fusion mechanisms have proven particularly effective for cross-modal matching problems, enabling fine-grained matching across heterogeneous embedding spaces. Our work adapts this paradigm to the CAD domain, where the two modalities are a B-Rep graph and a 2D engineering drawing, and the matching target is an individual B-Rep face rather than a complete shape.

1.2 Contributions

We address the problem of automatically associating 2D face indicators in PNG-based technical drawings with their corresponding faces in 3D CAD models, a core requirement for end-to-end PMI assignment in industrial manufacturing workflows. We introduce the task of self-supervised CAD-face localization: a point marker placed on a projected face in a 2D drawing must be matched to the correct face in the B-Rep graph. In a full deployment pipeline, this marker is supplied by detecting the tip of a PMI arrow in the drawing, making our approach a directly composable component of a complete PMI extraction and assignment system. Our specific contributions are:

1. A cross-modal transformer architecture combining a GATv2-based GNN for B-Rep geometry encoding with a ResNet-50 [12] or DINOv3 [13] image encoder for drawing region encoding, fused via cross-attention.
2. A self-supervised synthetic data generation pipeline producing paired STEP models and technical drawings at varying geometric complexity, with automatically derived ground-truth face-to-region correspondences requiring no manual annotation.
3. A systematic evaluation across datasets of varying complexity, including analysis of architectural variants and detailed error analysis, demonstrating that the model successfully learns cross-modal face correspondences.
4. A gap analysis identifying the key elements absent from current synthetic drawings relative to real industrial documents, providing a concrete roadmap toward deployment on real-world data.

2 Model Architecture

We propose a cross-modal architecture that learns to identify the 3D CAD face corresponding to a given point marker in a 2D technical drawing. The model consists of two encoding branches, one processing the 3D B-Rep graph, the other the 2D drawing crops, whose outputs are fused via cross-attention blocks and trained end-to-end with a bidirectional cross-entropy loss. An overview of the architecture is provided in Fig. 1.

2.1 3D Branch – GNN on B-Rep Geometry

The 3D branch operates directly on the boundary representation (B-Rep) of a CAD model provided as a STEP file. Rather than converting the geometry to a mesh or point cloud, we preserve the topological and geometric structure native to the CAD format by constructing a face-adjacency graph, where each node corresponds to a B-Rep face and

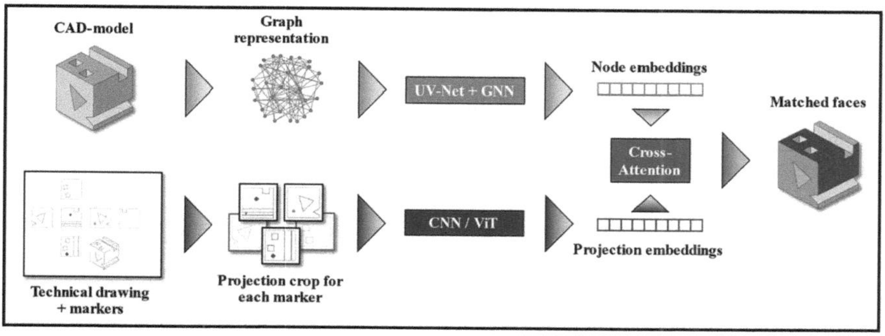

Fig. 1. Overview of the proposed cross-modal architecture for CAD-face localization.

each edge connects two faces that share a common boundary edge in the model topology. This representation naturally encodes the spatial relationships between surfaces without the geometric accuracy loss introduced by discretization.

Node Features. Each node is initialized with a feature vector composed of three complementary components. The first component consists of handcrafted geometric attributes encoding face-level properties including surface type, curvature type, orientation, face area, surface center, normal direction, center of mass, and UV parametric bounds. The second component is produced by UV-Net, an encoder pretrained for machining feature segmentation on the MFCAD dataset [14], where it learns to classify faces into 16 machining feature categories by processing each face in its parametric UV domain. This yields geometrically expressive per-face embeddings that capture fine-grained surface structure, including internal loops, holes, and boundary curves. The third component consists of Laplacian Positional Encodings (LPE) [15], derived from the eigenvectors of the graph Laplacian, which endow each node with an awareness of its structural position within the global B-Rep topology, disambiguating geometrically similar faces that occupy different positions in the model.

Edge Features. Each edge in the face-adjacency graph is initialized with a handcrafted feature vector encoding the geometry of the shared boundary curve between two adjacent faces. This includes the curve type, orientation, G-continuity class, convexity, length, 3D center position, parametric extent, and the dihedral angle between the two adjacent faces. These relational attributes provide the message passing operator with precise geometric context about the nature of the boundary shared between neighboring faces.

Message Passing. The initial node and edge features are refined through three layers of graph-based message passing using GATv2Conv with eight attention heads, a dynamic graph attention operator in which attention coefficients are conditioned jointly on both source and target node features. This allows each face to selectively aggregate information from its topological neighbors based on learned geometric relevance, rather than treating all neighbors equally. Through this process, each face representation is progressively enriched with contextual information from the surrounding geometry, producing a set of context-aware face embeddings that jointly capture local surface geometry and global topological structure.

2.2 2D Branch – Drawing Crop Encoder

The 2D branch processes the technical drawing in PNG-format to produce a set of projection embeddings, one per annotated face indicator. Given a technical drawing containing multiple orthographic views of the CAD model, each annotated projection is extracted as an individual image crop from the corresponding drawing. This yields a set of crop images, each containing one projection of the model.

To synthetically mark the target face within a crop, a point marker is placed at a representative location on the cropped projection. A single projection crop may contain multiple markers, one per visible annotated face. Rather than encoding these markers as hard pixel coordinates, they are represented as 2D Gaussian heatmaps centered at the marker positions, providing a smooth spatial signal that is tolerant to minor imprecision in marker placement. Each heatmap is stored as a dedicated second channel alongside the grayscale drawing crop, forming a two-channel input that jointly encodes drawing geometry and spatial face indication.

We evaluate two backbone configurations for encoding these crops. For ResNet-50, the standard three-channel first convolutional layer is replaced with a two-channel equivalent: both the drawing channel and the pointer channel are initialized from the first two channels of the pretrained RGB weights respectively, preserving as much pretrained knowledge as possible in the adapted input layer. For DINOv3 ViT-Small/16, a vision transformer backbone, the patch embedding projection is adapted to two channels by initializing the drawing channel from the mean of the pretrained RGB patch embedding weights and the pointer channel using Xavier initialization. In both configurations, all remaining pretrained weights are finetuned end-to-end during training. The ResNet-50 backbone produces a projection embedding via global average pooling over the final feature map, while DINOv3 uses the CLS token directly as the projection embedding without any additional pooling step.

2.3 Cross-Attention Fusion

To associate the 3D face embeddings with the 2D projection embeddings, the two sets of representations are fused through two stacked cross-attention blocks. In each block, node embeddings attend over the full set of drawing region embeddings, allowing each node to selectively aggregate information from the most relevant drawing regions. The output of each attention block is further refined through a feed-forward network with a residual connection, following standard transformer block design. Both the updated node embeddings and the (unchanged) region embeddings are L2-normalized prior to similarity computation, following the similarity scoring design of CLIP [9]. The matching scores are produced as dot products scaled by a learned temperature parameter, yielding a matrix of pairwise compatibility scores between all nodes and all annotated projections.

2.4 Training Objective

The model is trained using a bidirectional cross-entropy loss that enforces consistency in both matching directions simultaneously. Given the matrix of pairwise compatibility scores and the ground-truth face-to-region assignments, a cross-entropy loss is computed over all candidate nodes from the perspective of each annotated projection, identifying its correct node among all candidates and symmetrically over all candidate projections from the perspective of each matched face. Optimizing both directions jointly ensures that the mapping between face and region embeddings is consistent in both directions, preventing the model from finding degenerate solutions that score well in one direction only.

3 Synthetic Dataset Generation

A key challenge in training cross-modal models for CAD-drawing correspondence is the absence of large-scale annotated data pairing 3D CAD models with technical drawings and ground-truth face assignments. To overcome this, we develop a fully automated synthetic data generation pipeline that produces paired STEP models and technical drawings with precise face-level ground truth, requiring no manual annotation at any stage.

3.1 Parametric STEP Model Generation

All synthetic CAD models are generated programmatically as parametric box geometries with extruded features, using Open CASCADE Technology (OCCT) via the pythonOCC Python bindings.

The base geometry for our synthetic BoxFeat dataset is a box of randomized dimensions, onto which extruded features, rectangular pockets, cylindrical holes, and conical frustums, are placed procedurally at randomized positions and scales, with one to three features per box face. This yields models with three distinct surface types (planar, cylindrical, and conical) and no intersecting features, resulting in clean and unambiguous B-Rep topologies. The dataset comprises 4,342 models with an average of 38.21 B-Rep faces per model.

For broader evaluation, we additionally use the MFCAD dataset, an externally sourced benchmark of 15,488 machining feature models originally introduced for face-level segmentation. In contrast to our synthetic dataset, MFCAD models consist exclusively of planar faces but exhibit significantly higher topological complexity through intersecting machining features, representing an orthogonal axis of difficulty that complements the surface type diversity and higher face counts of BoxFeat. Representative examples of models from both configurations are shown in Fig. 2.

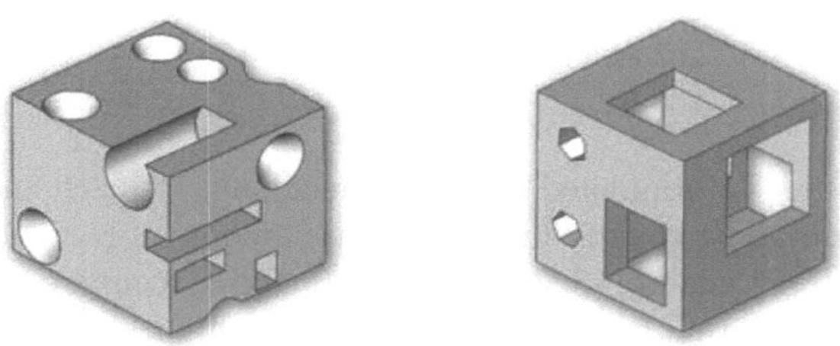

Fig. 2. STEP models from BoxFeat with open cylinder feature (left) and MFCAD with intersecting features (right).

3.2 Automated Technical Drawing Generation

Technical drawings are generated automatically from each STEP model using pythonOCC, which provides access to the hidden line removal (HLR) algorithm of Open CASCADE. For each model, standard orthographic views are computed and rendered as clean line drawings following standard first-angle projection conventions. HLR produces geometrically accurate projections, distinguishing visible from hidden edges and suppressing occluded geometry, resulting in drawings that closely resemble the projection content of manually authored technical drawings, currently without PMI annotations, dimension lines, or title blocks. Each view is extracted as an individual image crop, forming the set of projection crops used as an input to the 2D branch.

3.3 Ground Truth Face Mapping

In our synthetic data, it is straight-forward to exploit the link between CAD faces and projected areas to establish a ground truth face mapping. The marker that indicates the presence of PMI is placed such that it falls strictly within the interior of the projected face region. This excludes points that lie on the face boundaries, where projection overlap between adjacent faces would introduce label ambiguity.

4 Experiments

4.1 Experimental Setup

Datasets. We evaluate our method on two datasets summarized in Table 1. BoxFeat is our synthetic dataset of 4,342 procedurally generated CAD models. MFCAD is a large-scale external benchmark of 15,488 machining feature models included to evaluate performance beyond our synthetic domain. Both datasets are split into 90% training and 10% test sets. The model is trained independently on each dataset using an identical set of hyperparameters, which were selected based on BoxFeat performance and applied to MFCAD without further tuning.

Table 1. Dataset statistics. BoxFeat contains three surface types (76.0% planar, 8.7% cylindrical, 15.3% conical); MFCAD contains only planar faces.

Dataset	# Models	Avg. Faces (±std)	Range	# Surface Types	Intersections
BoxFeat	4,342	38.21 (±6.33)	18–60	3	No
MFCAD	15,488	22.62 (±5.30)	8–59	1	Yes

Evaluation. The task is formulated as N-way matching: for each annotated region in a drawing, the model must identify the correct face among all N candidate faces of the model. We report Top-1 and Top-3 accuracy, where Top-k accuracy counts a prediction as correct if the ground-truth face appears among the k highest-scored candidates. The random baseline is 1/N for Top-1 and 3/N for Top-3.

Variants. We compare two configurations of the 2D encoding branch: a ResNet-50 backbone pretrained on ImageNet and a DINOv3 ViT-Small/16 vision transformer pre-trained in a self-supervised manner on large-scale image collections. As described in Sect. 3.2 both variants share the same 3D branch and cross-attention fusion, allowing a controlled comparison of 2D feature extractors. Additionally, we evaluate ablated configurations to assess the contribution of individual architectural components.

4.2 Results

Table 2 reports Top-1 and Top-3 accuracy across all dataset and model variants. The random baseline reflects the difficulty of the task: uninformed assignment achieves only 2.62% Top-1 accuracy on BoxFeat and 4.42% on MFCAD, underscoring that high performance requires genuine cross-modal understanding. Both variants substantially outperform the random baselines. As can be seen, in all architectural variants and on both data sets, the Top-1 and Top-3 performances exceed 93% and 99% respectively. The DINOv3 variant outperforms the ResNet-50 variant, which is particularly visible in the Top-1 metric. DINOv3 achieves 98.04% $\pm$ 0.13% Top-1 accuracy across repeated runs on BoxFeat, confirming result stability. The gap between the Top-1 and Top-3 metrics suggests that a post-processing step might be beneficial, in which further arguments are collected to reprioritize top candidates. The results on MFCAD demonstrate strong generalization to an external dataset with a different geometric distribution, achieved without any hyperparameter retuning.

Table 2. Top-1 and Top-3 accuracy across datasets and model variants.

Dataset	BoxFeat		MFCAD	
Variant	Top-1	Top-3	Top-1	Top-3
ResNet-50	95.27%	99.66%	93.87%	99.77%
DINOv3	**98.04%**	**99.83%**	**96.54%**	**99.93%**
Random baseline	2.62%	7.85%	4.42%	13.26%

Table 3 reports the ablation results on BoxFeat. UV-Net contributes most substantially, with removal causing a drop of nearly 9 points which is almost identical to removing both UV-Net and LPE together, confirming that UV-Net is the dominant source of face disambiguation while LPE adds only marginal value. Notably, removing LPE on top of UV-Net improves over the UV-Net-only removal, suggesting that LPE provides limited complementary value without the richer UV-Net embeddings to build upon. GATv2 outperforms GATv1, validating the use of dynamic attention. Across GNN depths, performance peaks at a single layer and degrades with greater depth, consistent with over-smoothing; we retain three layers for robustness across varying model complexities. Notably, 0 message passing layers performs within 0.1 points of the full model, indicating that the initial node features are already highly expressive on synthetic data. Cross-attention fusion shows no consistent benefit on synthetic data, with the no-fusion variant outperforming the full model. We attribute this to the geometric fidelity of the synthetic drawings: clean line art directly reflects the underlying B-Rep geometry, allowing both encoders to independently converge to compatible embedding spaces without explicit cross-modal reasoning. We expect cross-attention to prove more beneficial on real industrial drawings, where visual clutter and annotation noise make the two modalities harder to align directly. The bidirectional loss provides a consistent advantage over the unidirectional variant, confirming that symmetric consistency constraints improve matching.

Table 3. Ablation study on BoxFeat using the DINOv3 backbone.

Variants	Top-1	Top-3
Full model	98.04%	99.83%
w/o UV-Net	89.25%	99.00%
w/o LPE	97.83%	99.75%
Base (no UV-Net, no LPE)	90.76%	99.29%
GATv1 instead of GATv2	97.36%	99.72%
0 GATv2 layers	97.93%	99.81%
1 GATv2 layer	**98.38%**	99.80%
2 GATv2 layers	98.13%	99.77%
4 GATv2 layers	97.61%	99.77%
0 cross-attn blocks	98.37%	**99.88%**
1 cross-attn block	97.93%	99.78%
Unidirectional loss	97.63%	99.81%

4.3 Discussion

The results demonstrate that the proposed architecture successfully learns cross-modal correspondences from purely synthetic training data, achieving near-perfect Top-3 accuracy on both datasets. The consistent advantage of DINOv3 over ResNet-50 suggests that transformer architecture, self-supervised pretraining, or both contribute to better feature transfer for the engineering drawing domain. Isolating these effects remains an interesting direction for future work. The high Top-3 accuracy across both variants suggests that the correct face is almost always among the top candidates, which is practically relevant: it shows the potential to establish near-perfect performance by implementing error correction and reprioritization loops. These loops may incorporate additional pattern matching models and a human-in-the-loop correction stage for ambiguous or ill-posed edge cases.

A more detailed analysis reveals that face area and graph size are no limiting factor, since Top-1 accuracy remains consistently high across faces of different sizes and models with varying numbers of faces, demonstrating that the architecture scales well to larger graphs without degradation. The most pronounced performance difference is observed across surface types. Planar and conical faces are recognized with high accuracy in BoxFeat, while cylindrical faces exhibit a notable drop. We attribute this to a specific geometric ambiguity: when a cylindrical feature is cut up to the boundary of the base box, the resulting cylinder is open on one side (see Fig. 2, left). When viewed from a lateral projection direction, such a partial cylinder presents a planar silhouette indistinguishable from a flat face, causing the model to confuse it with planar geometry. Such configurations are relatively rare in the dataset since cylinders are typically extruded within the box interior and therefore not visible in standard orthographic projections, excluding them from the annotated regions in most cases. Beyond surface type ambiguity, the remaining failure cases predominantly involve geometrically similar faces appearing at similar positions within the same projection crop, most notably in symmetric parts. Symmetric configurations represent a fundamentally harder challenge: when two or more faces are geometrically identical and symmetrically placed, no amount of local crop information can disambiguate them. Such cases are rare in BoxFeat by construction but more prevalent in MFCAD, and represent a primary contributor to remaining errors. Resolving them would require reasoning over the complete drawing structure rather than processing each crop in isolation and is a meaningful direction for future work.

As can be seen in Fig. 3, the synthetic drawings used for training and evaluation differ from real industrial technical drawings in several important respects. We now systematically identify these gaps, analyze their likely impact on model performance, and outline the extensions required to close them. This analysis serves both as an assessment of the current scope and as a concrete roadmap for future work.

The synthetic models in the current datasets are composed entirely of simple, analytically defined face types: planes, cylinders and cones. While these cover the most common face types in simple machined parts, real industrial CAD models frequently contain more complex surface geometry including freeform B-spline surfaces, blend fillets, and compound curved faces. Whether the learned embeddings generalize to complex surface types is an empirical question we leave for future work. Incorporating models with freeform and blended geometry, for example from the ABC dataset [16] into the training distribution is the most direct path toward addressing this gap, and represents a natural extension of the evaluation presented.

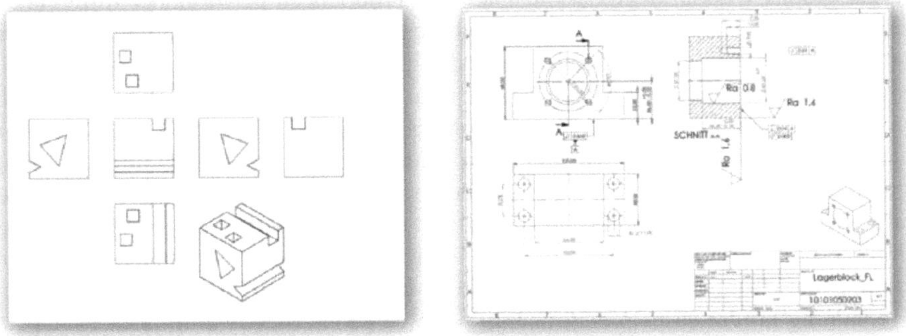

Fig. 3. Synthetic drawing from synthetic MFCAD-STEP-model (left) vs. real industrial drawing with PMI, title block, dimension lines, section views etc. (right) [1].

The current pipeline generates only standard orthographic projections following first-angle projection conventions. Real technical drawings frequently employ additional view types that are essential for communicating the geometry of complex parts: sectional views expose internal features such as blind holes and internal cavities that are otherwise invisible in standard projections; auxiliary views reveal true shapes of inclined faces; and detail views provide magnified depictions of small or intricate features. Each of these view types could be integrated in the synthetic data generation pipeline to introduce new geometric relationships between the 2D projection and the underlying 3D faces, requiring the model to reason about projection geometry beyond simple orthographic visibility.

In the current formulation, dot markers are placed exclusively on faces visible in a given projection view, with marker positions constrained to the interior of the projected face region. This excludes cases where the arrow tip points to the edge of a surface, for example side faces, which appear only as edges in standard orthographic views and carry no interior area in the projection. Addressing this limitation requires extending both the ground-truth face mapping and the marker placement strategy to handle edge-projected faces.

The synthetic drawings used in this work consist of pure orthographic projections containing only geometric outlines and edge lines, rendered without any additional content. Real industrial drawings, by contrast, are richly annotated documents containing PMI callouts, GD&T symbols, dimension lines, tolerance frames, surface finish indicators, title blocks, revision tables, and free-text notes. This additional content constitutes significant visual clutter relative to the geometric projection itself and introduces a distribution shift that the current 2D encoder, trained on clean crops, is not equipped to handle. Closing this gap requires augmenting the synthetic drawing generation pipeline to simulate realistic annotation content to make the 2D encoder robust to drawing clutter, as well as a finetuning step on carefully annotated real-world data.

5 Conclusion

We presented a cross-modal transformer architecture for self-supervised CAD-face localization, addressing the problem of automatically associating 2D face indicators in technical drawings with their corresponding faces in 3D CAD models. By formulating the task as N-way matching between drawing region embeddings and B-Rep face embeddings, and training exclusively on synthetically generated data with automatically derived ground-truth labels, we eliminate the need for manual annotation entirely. Our architecture combines a GATv2-based graph neural network encoding B-Rep topology and geometry with a ResNet-50 or DINOv3 backbone encoding 2D drawing, fused via a cross-attention mechanism and optimized with a bidirectional cross-entropy loss. On the BoxFeat dataset our best configuration achieves 98.04% Top-1 accuracy, far exceeding the random baseline of 2.62% and demonstrating that genuine cross-modal geometric reasoning is learned from synthetic data alone. Applied to MFCAD the same model achieves 96.54% accuracy without any hyperparameter retuning, demonstrating strong generalization across B-Rep distributions. The gap analysis in Sect. 4.3 identifies four concrete limitations of the current approach relative to real industrial drawings: the limited diversity of face geometry, the restriction to standard orthographic projections, the exclusion of edge-projected and partially visible faces, and the absence of drawing clutter and PMI content. Each of these represents a targeted extension of the synthetic data generation pipeline rather than a fundamental architectural limitation, and we regard closing these gaps as the primary direction for future work. Combined with domain adaptation toward real industrial drawings, we expect the proposed architecture to form a viable foundation for fully automated, drawing-driven PMI population in industrial CAD environments.

6 Disclosure of Interests.

The authors have no competing interests to declare that are relevant to the content of this article.

Acknowledgments. This work was carried out in collaboration with PartSpace GmbH within the project CADistency (01/2025–12/2027), funded by the Bavarian State Ministry of Economic Affairs, Regional Development and Energy (StMWi) under the Digitalization funding line of the Bavarian Cooperative Research Program. PartSpace GmbH contributed parts of the technical foundation of the work, including software architecture and code.

References

1. Grieshammer, M., Kellner, P., Götz, M., Schwarz, A., Ulrich, D., Schirmeier, F.: KI als Brücke zwischen technischer Zeichnung und CAD-Modell. Masch. Bau **5**, 10–13 (2025). https://doi.org/10.1007/s44029-025-1318-5
2. Jayaraman, P.K., Sanghi, A., Lambourne, J.G., et al.: UV-net: learning from boundary representations. In: Proceedings of the IEEE/CVF Conference on Computer Vision and Pattern Recognition (CVPR), pp. 11703–11712. IEEE (2021)

3. Lambourne, J.G., Willis, K.D.D., Jayaraman, P.K., et al.: BRepNet: a topological message passing system for solid models. In: Proceedings of the IEEE/CVF Conference on Computer Vision and Pattern Recognition (CVPR), pp. 12773–12782. IEEE (2021)
4. Colligan, A.R., Robinson, T.T., Nolan, D.C., et al.: Hierarchical CADNet: learning from B-Reps for machining feature recognition. Comput. Aided Des. **147**, 103226 (2022). https://doi.org/10.1016/j.cad.2022.103226
5. Brody, S., Alon, U., Yahav, E.: how attentive are graph attention networks? In: International Conference on Learning Representations (ICLR) (2022). https://openreview.net/forum?id=F72ximsx7C1
6. Jamieson, L., Moreno-García, C.F., Elyan, E.: A review of deep learning methods for digitisation of complex documents and engineering diagrams. Artif. Intell. Rev. **57**(6), 136 (2024). https://doi.org/10.1007/s10462-024-10779-2
7. Khan, M.T., Yong, Z., Chen, L., et al.: Automated parsing of engineering drawings for structured information extraction using a fine-tuned document understanding transformer. In: IEEE International Conference on Industrial Engineering and Engineering Management (IEEM) (2025). https://doi.org/10.48550/arXiv.2505.01530
8. Wang, X., Zheng, J., Hu, Y., et al.: From 2D CAD drawings to 3D parametric models: a vision-language approach. In: Proceedings of the AAAI Conference on Artificial Intelligence (AAAI 2025). https://doi.org/10.48550/arXiv.2412.11892
9. Radford, A., Kim, J.W., Hallacy, C., et al.: Learning transferable visual models from natural language supervision. In: Proceedings of the 38th International Conference on Machine Learning (ICML), vol. 139, pp. 8748–8763. PMLR (2021)
10. Wang, F., Kang, L., Li, Y.: Sketch-Based 3D Shape Retrieval using Convolutional Neural Networks. arXiv:1504.03504 (2015)
11. Qin, J., Yuan, S., Chen, J., et al.: SHREC'22 track: sketch-based 3D shape retrieval in the wild. Comput. Graph. **107** (2022). arXiv:2207.04945
12. He, K., Zhang, X., Ren, S., Sun, J.: Deep residual learning for image recognition. In: Proceedings of the IEEE/CVF Conference on Computer Vision and Pattern Recognition (CVPR), pp. 770–778. IEEE (2016)
13. Siméoni, O., Vo, H.V., Seitzer, M., et al.: DINOv3. arXiv preprint arXiv:2508.10104 (2025). https://arxiv.org/abs/2508.10104
14. Cao, W., Robinson, T., Hua, Y., et al.: Graph representation of 3D CAD models for machining feature recognition with deep learning. In: Proceedings of the ASME International Design Engineering Technical Conferences (IDETC-CIE). ASME (2020)
15. Dwivedi, V.P., Joshi, C.K., Luu, A.T., et al.: Graph neural networks with learnable structural and positional representations. In: International Conference on Learning Representations (ICLR) (2022). https://arxiv.org/abs/2110.07875
16. Koch, S., Matveev, A., Jiang, Z., et al.: ABC: A big CAD model dataset for geometric deep learning. In: Proceedings of the IEEE/CVF Conference on Computer Vision and Pattern Recognition (CVPR), pp. 9601–9611. IEEE (2019)

Challenges in Multimodal Fusion of LiDAR and Camera Data for Road Surface Analysis

Simon Hardt(✉), Matthias Micheler, Gesine Wagner, and Ulrich Göhner

Kempten University of Applied Sciences, Kempten, Germany
{simon.hardt,matthias.micheler,ulrich.goehner}@hs-kempten.de

Abstract. Accurate road condition assessment requires the fusion of geometric and visual information from LiDAR and camera systems. In this work, we investigate the feasibility of multimodal registration between LiDAR-derived height and intensity maps and RGB image data for road surface analysis under real-world conditions.

Our experiments show that classical feature- and intensity-based registration methods are not suitable for this application. The homogeneous nature of road surfaces, the lack of stable cross-modal correspondences, and the limited structural quality of LiDAR intensity data prevent robust and reproducible alignment.

As a result, we rely on a time- and velocity-based alignment strategy that leverages sensor synchronization and vehicle motion. This approach provides a stable coarse registration but does not allow precise multimodal fusion and is sensitive to temporal offsets and dynamic measurement conditions.

These findings indicate that reliable multimodal fusion cannot currently be achieved using classical methods in real-world road scenarios. We therefore outline learning-based approaches, such as cross-modal feature learning and self-supervised methods, as promising directions for future work.

Keywords: multimodal fusion · LiDAR · camera · registration · road analysis

1 Introduction

As road networks expand and traffic demand increases, maintenance costs continue to rise. Consequently, automated road condition assessment is gaining importance as a cost-efficient solution, as it enables predictive maintenance strategies that facilitate early damage detection and help prevent costly large-scale repairs. We present a measurement system designed for deployment across various vehicle types. In particular, the system is intended for integration into regularly operating vehicles, such as buses, to enable continuous road condition monitoring. The system combines a solid-state LiDAR sensor with a high-resolution RGB camera. The LiDAR system provides georeferenced height and

© The Author(s), under exclusive license to Springer Nature Switzerland AG 2027
M. Geiß et al. (Eds.): DEXA 2026 Workshops, CCIS 3078, pp. 137–142, 2027.
https://doi.org/10.1007/978-3-032-34202-7_14

reflection-intensity data, while the RGB camera is used to identify relevant road surface regions. This enables filtering of the LiDAR data so that it can be specifically applied for accurate road surface measurement.

In this work, we investigate multimodal image registration under real-world conditions, focusing on the challenges posed by low feature density and limited cross-modal feature visibility.

2 Problem Statement

The multimodal registration of LiDAR and camera data in this application is inherently challenging due to the characteristics of the data and the underlying sensor principles.

A central issue is the homogeneous nature of the road surface, which provides only limited distinctive structures. As a result, classical feature-based computer vision methods fail to extract stable keypoints and establish reliable correspondences between modalities [2]. Furthermore, LiDAR and camera data differ significantly in their representation. The LiDAR system incrementally reconstructs the road surface during motion and generates a bitmap image containing height and intensity information, with a resolution of $10mm/pixel$(see Fig. 1 b and c). Whereas the camera captures image strips along the driving direction (see Fig. 2 a), complicating transformation into a common coordinate system. In addition, LiDAR intensity data often exhibits weak structural information, further limiting cross-modal comparability.

Overall, this results in a lack of stable cross-modal features, leading to unstable and inaccurate alignments. Therefore, alignment is performed using temporal synchronization and vehicle motion. While this enables coarse registration, it is prone to drift and cannot be applied consistently over longer sequences, preventing precise multimodal fusion.

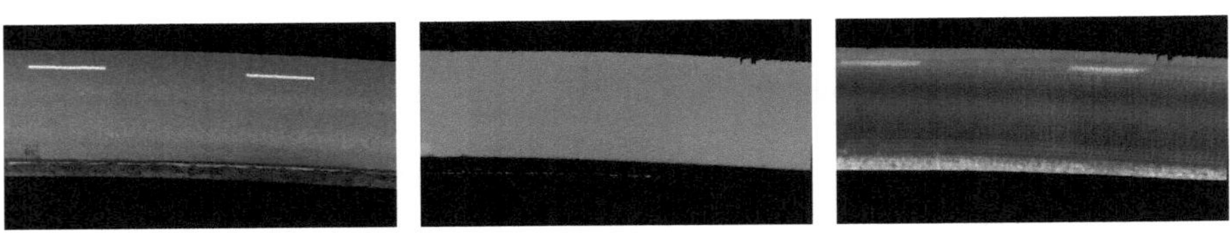

(a) Fused camera frames (b) Height (c) Intensity

Fig. 1. Comparison of stitched camera, LiDAR height, and LiDAR intensity data

3 Multimodal Fusion Approach

Our data fusion algorithm follows a coarse-to-fine approach. In the coarse registration stage, the degrees of freedom of the transformation are reduced to simplify the alignment problem. The resulting rigid transformation T is therefore parameterized only by translations in the x- and y-directions and a rotation θ around the image center.

$$T = \begin{bmatrix} \cos\theta & -\sin\theta & x \\ \sin\theta & \cos\theta & y \\ 0 & 0 & 1 \end{bmatrix} \tag{1}$$

The transformation T defines the spatial mapping from the LiDAR-derived bitmap to the camera image undergoing registration. The LiDAR data is already available as a height map in a top-down perspective (xy-Plane). For the RGB camera, a homographic projection is computed during calibration to transform the images into a corresponding top-down view and resolution (see Fig. 2 (b)).

To further reduce the space of possible alignments, we use sensor timestamps from the RGB images and associate them with the trajectory provided by the LiDAR system. Since the LiDAR sensor and the camera have a fixed spatial relationship, we can estimate T that places each camera frame relative to the corresponding point along the vehicle trajectory. The vehicle trajectory is represented by the sequence of vehicle poses sampled at LiDAR frame acquisition times, expressed in the LiDAR bitmap coordinate frame.

To mitigate errors caused by the lack of temporal synchronization between LiDAR and camera acquisition, the transformation $T(t)$ at the camera-frame timestamp t is estimated by interpolating between preceding T_1 and subsequent T_2 vehicle poses obtained from LiDAR frames.

$$T(t) = \frac{t_2 - t}{t_2 - t_1} T_1 + \frac{t - t_1}{t_2 - t_1} T_2 \tag{2}$$

The described coarse registration approach provides a good initial alignment. Under favorable conditions, such as planar road surfaces with minimal elevation variation, fine registration may be omitted. In Fig. 1 (a), the result of a coarse registration is shown.

To address challenging scenarios, reduce the need for manual calibration, and mitigate temporal misalignment between LiDAR and camera data, we evaluated various registration methods for multimodal image registration and fusion.

This includes the registration between LiDAR-derived maps (height and intensity) and RGB camera frames. Specifically, we investigated the registration of the intensity map to the camera images as well as the height map to the camera images. For the registration, we assume that the images share the same perspective and scale. Therefore, only planar transformations—translation in x and y, and rotation about the z-axis—need to be estimated.

We focus on classical computer vision approaches for multimodal image registration [6]. To quantify the results and enable the use of intensity-based methods, we compared different similarity metrics. For this purpose, the camera frames are downsampled to match the resolution of the LiDAR data, and the metric responses are analyzed. As ground truth, we use manually registered frames defining the reference transformation. We evaluate the similarity metric in a local neighborhood around this solution by varying x and y within ±2 units, expecting a peak at the ground-truth alignment. The most promising results are obtained using Mutual Information (MI) [6] and the Structural Similarity Index (SSIM) [7].

It is noteworthy that these metrics can yield high similarity scores even for severely misaligned images. This is primarily due to the characteristics of road scenes, where large portions of the image consist of homogeneous surfaces with little structural information, limiting the discriminative power across modalities. To address this issue and make the metrics applicable in the road context, we restrict the analysis to regions of interest (ROIs) containing distinctive features such as lane markings, manhole covers, potholes, and other structured elements. The similarity metrics are then computed exclusively within these regions. In our experiments, the metrics distinguish between coarse registrations with misaligned and correctly synchronized timestamps over an entire recording session. However, when applied in intensity-based registration, the metrics often produce higher similarity scores for slightly misaligned transformations.

We also evaluated different feature extractors, such as ORB [5], as well as a variant adapted for multimodal images (Locally Normalized Image Feature Transform (LNIFT)) [3]. However, no configuration reliably extracted corresponding features from LiDAR and camera data, making feature matching impractical.

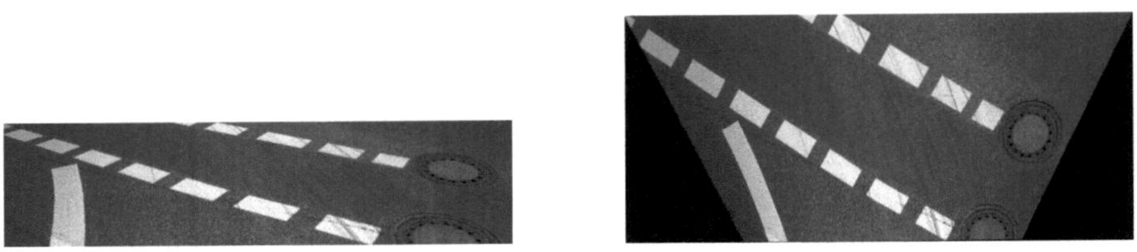

(a) Captured camera frame. (b) Preprocessd camera frame.

Fig. 2. Exeample preprocesseiong from RGB-Camera frames.

4 Limitations of Multimodal Fusion

Our multimodal fusion approach, which relies solely on coarse alignment-based registration, yields good results under favorable conditions. However, performance degrades noticeably in the presence of timestamp misalignment, in which case manual alignment was often required. In addition, the homographic projection of the camera frames into a top-down perspective is only accurate when the vehicle is approximately level. When the vehicle experiences roll or pitch—for example, when driving on uneven terrain or slopes—projection errors become apparent and negatively affect the coarse registration results.

In our evaluation of multimodal fusion, the main challenges originated from the LiDAR datasets. First, the spatial resolution of the LiDAR data 10 mm px^{-1} is significantly lower than that of the camera images. In combination with the limited vertical (Z) resolution of the LiDAR system, small-scale features such as cracks and minor potholes cannot be reliably captured and are therefore not suitable for registration. Additionally, the LiDAR data exhibits noise-like pre-processing artifacts and inconsistencies in the representation of features, most notably along linear structures such as road markings. These effects lead to unstable feature detection and reduced reliability of the evaluated similarity metrics. The proposed ROI-based metric approach improves discriminative performance in regions with a high density of distinctive features. However, its effectiveness is limited in road sections with sparse or absent structural features.

5 Discussion

Classical registration methods are limited, as homogeneous road surfaces do not provide stable keypoints or reliable correspondences between LiDAR and camera data. This is also reflected in the evaluation metrics, where SSIM [7] values of 0 occur in feature-sparse regions, indicating the absence of usable structural correspondences.

Against this background, the alignment of the datasets was performed using timestamps and vehicle motion, but this approach is subject to several limitations. Even a temporal offset of 20 s leads to significantly degraded metric values, highlighting the high sensitivity to synchronization errors. In addition, inaccuracies in velocity estimation as well as dynamic effects such as vibrations and changes in vehicle inclination introduce further registration errors [2].

While the coarse registration yields overall stable results, deviations of up to approximately 1 m between LiDAR and camera data occur in more complex scenarios. Furthermore, strong height variations in the road surface lead to local misalignments, as the underlying model assumptions are only partially fulfilled.

Overall, a clear trade-off between robustness and accuracy can be observed: while the sensor-based alignment provides stable results, the achievable precision remains insufficient for accurate multimodal fusion.

6 Conclusion and Future Work

Multimodal fusion is currently limited to coarse pre registration, as precise alignment is hindered by missing cross modal correspondences. Time and velocity based alignment provides a suitable initialization for further refinement. Building on this, learning-based approaches can either improve the alignment by refining the coarse registration or utilize the pre-aligned data as a basis for training and evaluation.

Recent work shows that deep neural networks can learn cross modal correspondences directly from data and outperform classical techniques in challenging environments [1].

Self-supervised approaches further reduce the dependency on labeled data by learning modality invariant representations [8]. In addition, transformer-based and detection-free methods improve robustness by learning dense correspondences instead of relying on explicit keypoints [4].

Overall, AI-based approaches are a key enabler for robust multimodal fusion in road condition assessment.

References

1. Cattaneo, D., Valada, A.: CMRnext: camera to lidar matching in the wild for localization and extrinsic calibration (2025). https://arxiv.org/abs/2402.00129
2. Hardt, S.: Registrierung von 2D- und 3D-Datensätzen mittels multimodaler Methoden zur Fusion homogener Oberflächen. Master's thesis, Hochschule Kempten (2025). unpublished
3. Li, J., Xu, W., Shi, P., Zhang, Y., Hu, Q.: Lnift: locally normalized image for rotation invariant multimodal feature matching. IEEE Trans. Geosci. Remote Sens. **60**, 1–14 (2022). https://doi.org/10.1109/TGRS.2022.3165940
4. Li, M., et al.: 2D3D-matr: 2D-3D matching transformer for detection-free registration between images and point clouds (2023). https://arxiv.org/abs/2308.05667
5. Rublee, E., Rabaud, V., Konolige, K., Bradski, G.: Orb: an efficient alternative to sift or surf. In: 2011 International Conference on Computer Vision, pp. 2564–2571 (2011). https://doi.org/10.1109/ICCV.2011.6126544
6. Velesaca, H.O., Bastidas, G., Rouhani, M., Sappa, A.D.: Multimodal image registration techniques: a comprehensive survey. Multimedia Tools Appl. **83**(23), 63919–63947 (2024). https://doi.org/10.1007/s11042-023-17991-2
7. Wang, Z., Bovik, A., Sheikh, H., Simoncelli, E.: Image quality assessment: from error visibility to structural similarity. IEEE Trans. Image Process. **13**(4), 600–612 (2004). https://doi.org/10.1109/TIP.2003.819861
8. Zhang, Y., Hou, J., Ren, S., Wu, J., Yuan, Y., Shi, G.: Self-supervised learning of lidar 3D point clouds via 2D–3D neural calibration. IEEE Trans. Pattern Anal. Mach. Intell. **47**(10), 9201–9216 (2025). https://doi.org/10.1109/tpami.2025.3584625

Author Index

© The Editor(s) (if applicable) and The Author(s), under exclusive license
to Springer Nature Switzerland AG 2027

M. Geiß et al. (Eds.): DEXA 2026 Workshops, CCIS 3078, pp. 143–144, 2027.
https://doi.org/10.1007/978-3-032-34202-7

MIX
Papier aus verantwortungsvollen Quellen
Paper from responsible sources
FSC® C105338

If you have any concerns about our products,
you can contact us on
ProductSafety@springernature.com

In case Publisher is established outside the EU,
the EU authorized representative is:
Springer Nature Customer Service Center GmbH
Europaplatz 3, 69115 Heidelberg, Germany

Printed by Libri Plureos GmbH
in Hamburg, Germany